ADVANCES IN
DNA SEQUENCE-SPECIFIC AGENTS

Volume 4 • 2002

ADVANCES IN
DNA SEQUENCE-SPECIFIC AGENTS

Series Editor: GRAHAM B. JONES
Department of Chemistry
Northeastern University
Boston, Massachusetts

Volume Editor: BRANT J. CHAPMAN
Department of Chemistry
Montclair State University
Upper Montclair, New Jersey

VOLUME 4   •   2002

ELSEVIER

Elsevier
Amsterdam – Boston – London – New York – Oxford – Paris
San Diego – San Francisco – Singapore – Sydney – Tokyo

ELSEVIER SCIENCE B.V.
Sara Burgerhartstraat 25
P.O. Box 211, 1000 AE Amsterdam, The Netherlands

First edition 2002

Library of Congress Cataloging in Publication Data
A catalog record from the Library of Congress has been applied for.

ISBN: 0–444–51096–6
ISSN: 1067–568X (Series)

Transferred to digital printing 2006
Printed and bound by CPI Antony Rowe, Eastbourne

CONTENTS

LIST OF CONTRIBUTORS

Giuseppe Bifulco
Dipartimento di Scienze Farmaceutiche
Università di Salerno
Italy

Michael Brenowitz
Department of Biochemistry
Albert Einstein College of Medicine
New York

Walter J. Chazin
Department of Biochemistry & Center for Structural Biology
Vanderbilt University
Nashville, Tennessee

Michael E. Colvin
Biology and Biotechnology Research Program
Division of Molecular and Structural Biology
Lawrence Livermore National Laboratory
Livermore, California

Gauri M. Dhavan
Department of Biochemistry
Albert Einstein College of Medicine
New York

Irving H. Goldberg
Department of Biological Chemistry and Molecular
    Pharmacology
Harvard Medical School
Boston, Massachusetts

Luigi Gomez-Paloma
Dipartimento di Scienze Farmaceutiche
Università di Salerno
Italy

Keisuke Makino
Institute of Advanced Energy
Kyoto University
Japan

A.K.M.M. Mollah
Department of Biochemistry
Albert Einstein College of Medicine
New York

Takashi Morii
Institute of Advanced Energy
Kyoto University
Japan

Judy N. Quong   Chemistry and Materials Science Directorate
                               Lawrence Livermore National Laboratory
                               Livermore, California

Jarrod A. Smith   Department of Biochemistry & Center for Structural Biology
                               Vanderbilt University
                               Nashville, Tennesee

Jan M. Woynarowski   Cancer Therapy and Research Center and Department
                                 of Radiation Oncology
                               University of Texas Health Science Center
                               San Antonio, Texas

Zhen Xi   Department of Biological Chemistry and Molecular
                                 Pharmacology
                               Harvard Medical School
                               Boston, Massachusetts

Preface

DNA sequence specificity plays a critical role in a number of biological processes, and influences a diverse range of molecular recognition phenomena, including protein–DNA, oligomer–DNA, and ligand–DNA interactions. This series will give the reader an up-to-date view of both established and emergent trends, in research involving DNA-interactive agents with an emphasis on sequence specificity. The series encompasses design, synthesis, application, and analytical methods (including clinical and *in vitro*) for the study of these critical interactions. As our understanding of the genome and proteome expands, general developments in the field of DNA sequence-specific interactions are likely to play an increasingly important role. Key developments are likely to involve small molecules with highly specific groove binding, molecules with affinity for unique DNA microenvironments, oligonucleotides capable of stabilizing unique DNA architectures, macromolecules which form unique ternary complexes, and molecules which influence the recruitment and assembly of transcription factors. Accordingly, manuscripts have been solicited from experts covering a diverse range of fields, reflecting the cross-disciplinary and dynamic nature of the series. Volume 4 of this series describes work on the modification of DNA by AT-specific anticancer drugs (Chapter 1), DNA alkylation events which involve metabolite generation (Chapter 2), DNA sequence recognition by two selective binders (Chapter 3), bulged DNA microenvironments as molecular targets (Chapter 4), DNA sequence-specific binding by short peptides (Chapter 5) and the analysis of DNA–protein interactions using DNase I footprinting methodology (Chapter 6).

As the ramifications of genome decoding become apparent, we enter a period of enormous opportunity for chemists and molecular biologists alike. In this regard we are confident that the reviews herein will serve to stimulate new ideas, as the boundaries of what has become a vibrant and dynamic field, continue to be extended.

Graham B. Jones
Series Editor

Brant J. Chapman
Volume Editor

ELSEVIER

Advances in DNA Sequence-specific Agents 04
(2002) 1–27

Advances
in DNA
Sequence-specific
Agents

www.elsevier.nl/locate/series/adna

Preferential damage to defined regions of genomic DNA by AT-specific anticancer drugs

Jan M. Woynarowski*

Cancer Therapy and Research Center and Department of Radiation Oncology, University of Texas Health Science Center, San Antonio, TX 78245, USA

1. Introduction

A. Targeting critical regions of the genomic DNA

DNA lesions localized in various domains of the genome are likely to differ in their biological consequences. Targeting of domains in cellular DNA, which are crucial for tumor cell functioning, is the intuitive prerequisite for a region specific strategy [1–4]. Many currently used DNA-reactive antitumor drugs need to form extensive lesions in cellular DNA to inhibit cell proliferation. Sometimes, however, damage to bulk cellular DNA does not show a consistent relationship with cytotoxic effects, but damage to specific regions does [5,6]. It is possible that only a few hits may be sufficient for inhibition of tumor cell growth, provided these lesions are of the right type and are located in regions critical for cancer cell proliferation.

Region-specific DNA damage offers, in principle, a potential to increase treatment selectivity for tumor versus normal cells. In contrast to critically located lesions, damage to unessential targets may contribute little to the drug's overall ability to kill cancer cells, yet it is likely to result in toxic and mutagenic side effects for normal cells. Thus, even specific drugs that are capable of damaging crucial targets may form a vast majority of lesions elsewhere, causing mainly "collateral damage" to unessential or less essential targets. Elimination, or at least reduction, of this unessential damage may have significant therapeutic benefits and is a necessary aspect to consider in exploring region-specific strategies.

* Corresponding author.
E-mail address: jmw1@ saci.org
Advances in DNA Sequence-specific Agents, Volume 4, 1–27

Until recent years, the mainstream approach for region specificity utilized triple-helix-forming oligonucleotides, which can achieve very high sequence specificity in cell-free systems [7]. Oligonucleotides have a potential to specifically recognize a single site in the genome. However, despite the conceptually inspiring findings in cell-free systems, triple-helix-forming oligonucleotides are only poorly to modestly growth inhibitory to cultured tumor cells *in vitro* and have no reported antitumor *in vivo* activity. This lack of activity reflects a combination of adverse factors, such as restrictions on permitted target sequences, inadequacy of intracellular conditions (e.g. sub-optimal pH or Mg^{2+} levels), and poor oligonucleotide uptake and their intracellular degradation [8–11].

Recent progress in rational design of small molecules has resulted in the ability to predictably modulate drug-DNA sequence recognition [12–16]. Especially promising are advances in sequence-specific recognition of DNA motifs in 2:1 complexes formed in the minor groove by polyamide compounds containing pyrrole and imidazole moieties [4,12,17–19]. Such polyamides, containing additional alkylating groups, were reported to recognize predetermined motifs of several base pairs (bp) in short model sequences of naked DNA with affinities and specificities comparable to those of DNA-binding proteins [4,12,20].

Despite the advances in sequence specificity at the level of short oligonucleotide tracts, mainly monitored in model cell-free systems, the potential of small molecular weight agents for localized damage in longer stretches of DNA (i.e. region specificity) remains largely unexplored. Region specificity implies that an agent is able to specifically recognize a motif within the intended target region, but also that such a motif is absent from, or scarce, elsewhere in the genome. A specificity of a single hit per genome requires agents recognizing discrete motifs of 15–16 bp [3]. To preserve a drug's ability to penetrate cancer cells, currently available small molecules are designed to recognize shorter motifs (<10 bp). The caveat is that such motifs are likely to be found in a variety of loci other than their intended targets.

A few experimental and/or computational studies, did address the question of which regions of genomic DNA can be preferentially affected by drugs of various sequence preferences at the nucleotide level [2,21–27]. The studies on region specificity of small molecular weight anticancer drugs, reviewed in this chapter, were mainly focused on several types of AT-specific drugs. However, the established generalized determinants of region specificity and the approach to assessing drug potential for damage to defined regions of the genome can be extrapolated to other agents as well. The reviewed examples indicate that region specificity of DNA damage is not restricted to the "one hit per genome" scenario. Certain AT-sequence-specific small molecules have been demonstrated to produce a non-random, region-specific damage to genomic DNA that could well prove to be an important factor in their biological effects, including antitumor activities.

B. AT-specific small molecular weight drugs and their potential targets

AT-specific anticancer drugs   The new generations of DNA-reactive drugs can interact with DNA in a remarkably specific way. For a number of such drugs, high DNA binding specificity is accompanied by significant cytotoxic and antitumor activities. This is particularly true for AT-specific compounds "reading" a DNA sequence from the minor groove.

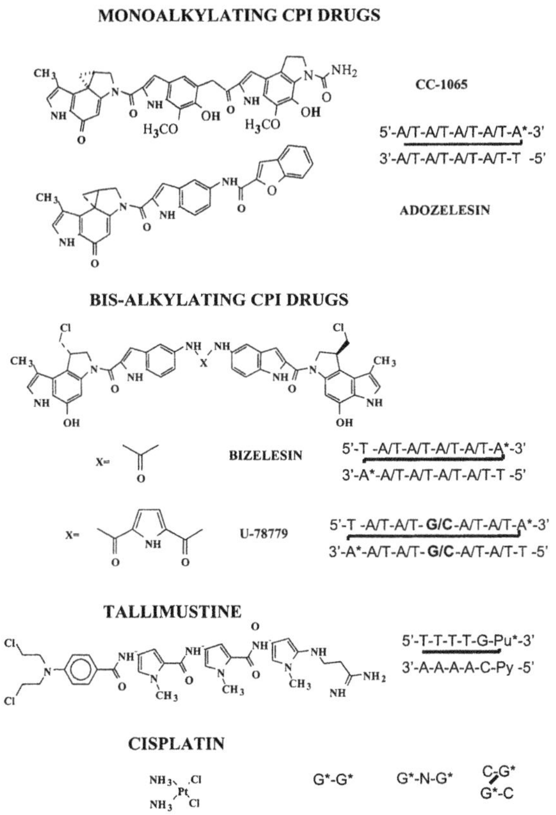

MONOALKYLATING CPI DRUGS

CC-1065

5'-A/T-A/T-A/T-A/T-A*-3'
3'-A/T-A/T-A/T-A/T-T -5'

ADOZELESIN

BIS-ALKYLATING CPI DRUGS

X= BIZELESIN

5'-T -A/T-A/T-A/T-A/T-A*-3'
3'-A*-A/T-A/T-A/T-A/T-T -5'

X= U-78779

5'-T -A/T-A/T- G/C-A/T-A/T-A*-3'
3'-A*-A/T-A/T- G/C-A/T-A/T-T -5'

TALLIMUSTINE

5'-T-T-T-T-G-Pu*-3'
3'-A-A-A-A-C-Py -5'

CISPLATIN

G*-G*   G*-N-G*   C⌒G*
                                G*-C

Fig. 1. Examples of highly cytotoxic and *in vivo* antitumor AT-specific anticancer drugs. DNA binding motifs shown for these drugs are typical rather than comprehensive (see details in text). Asterisks in these motifs indicate adducted DNA bases. Cisplatin is included as an example of G-alkylating agent of low sequence specificity.

Examples of highly cytotoxic and AT-specific drugs are given in Figure 1 along with their typical or preferred binding motifs.

Drugs of the well characterized cyclopropylpyrroloindole (CPI) group (CC-1065, adozelesin, bizelesin) bind preferentially to runs of 5–7 ATs and alkylate adenine residues at the 3' ends [28–33]. CPI drugs are among the most cytotoxic compounds ever discovered, inhibiting cell growth in the picomolar range of concentrations. For example, drug concentrations producing 50% cell growth inhibition (GI_{50}) for bizelesin and adozelesin in human leukemic CEM cells amount to 0.2–0.6 and 14 pM, respectively [24,25]. Some CPI analogs were designed to target mixed A/T-G/C sequences (e.g. U-78779, Fig. 1) [26,34]. The parent natural compound CC-1065 has been eliminated from further development due to severe toxic effects. Adozelesin and bizelesin, however, are in clinical trials [35,36] and synthetic efforts continue to obtain various new CPI analogs [37,38].

Tallimustine (FCE 24517), another highly AT-specific drug, is an alkylating distamycin analog in clinical trials as a broad-spectrum anticancer drug [39–42]. Tallimustine exemplifies the trend among several groups to use distamycin/netropsin N-methylpyrrole-based polyamide structures as a core to design sequence-specific agents (for review see [14,43,44]).

Drug-DNA adducts are not the only type of highly cytotoxic AT-specific DNA lesions. Enediyne compound C-1027, is an example of an AT-specific strand scission agent [45]. C-1027 forms double strand breaks preferentially in the runs of 4–5 A/T [45]. The ability of C-1027 to induce localized strand breakage has been demonstrated in intracellular AT-rich regions of SV40 DNA [46] and in the origin of replication of an intracellular episomal DNA [47].

In contrast to the AT-recognizing agents, "classical" DNA-reactive drugs with *in vivo* antitumor activity tend to exhibit relatively low sequence specificity at the nucleotide level. For instance, cisplatin and other platinum complexes, and various other alkylating drugs, including some simple and composite analogs of nitrogen mustards, recognize short simple motifs such as GG or GC [1,2,23,48–50]. Whereas several clinically useful drugs are among such rather non-specific agents, they are usually markedly less potent on a molar basis than AT-specific compounds.

Potential AT-rich target regions   It is intriguing that so many highly cytotoxic and anti-tumor *in vivo* drugs of divergent structural and mechanistic features share a preference for AT sites. Can targeting of AT-rich domains be more lethal to the cell than targeting random or G/C-rich sequences? AT-rich areas are often identified within functionally important regions, but are all types of AT-rich domains equally critical? Matrix Associated Regions (MARs), which play a key role in the organization of nuclear chromatin, are among the most crucial kinds of AT-rich loci. MAR domains anchor the loops of genomic DNA on the nuclear matrix [51–54] and often contain, or are in proximity to, other specific AT regulatory elements, e.g. DNA Unwinding Element (DUE) in the origins of replica-tion, TATA box in transcription initiation sites, topoisomerase II and SATB1 binding sites [51,55–57]. Such specialized AT-rich loci are involved in local modifications of DNA/chromatin structure (dynamic structural polymorphism, e.g. local helix destabilization) that play an essential role in chromatin organization, the initiation of replication and tran-scription, and mitosis [52,53,58]. Thus, damage to such AT elements by anticancer DNA-reactive drugs is likely to impede the regulation of the processes that are crucial for cell proliferation. AT-specific drugs may help to uncover which of the AT-rich regions/elements can have a therapeutic utility.

C. Approach to establishing drug region specificity

A systematic approach to characterize region specificity of DNA damage [23–27,33] can comprise the following elements: (i) experimental determination of drug sequence specificity (binding motifs), (ii) analysis of motif distribution in a representative selec-tion of sequences to predict which regions may be particularly vulnerable, and (iii) the experimental verification of such predictions by quantitating drug-induced lesions in selected specific regions. The general outline of this approach is given below and specific examples are described in the subsequent sections.

Sequence specificity   The knowledge of drug binding motif(s) at the nucleotide level is the necessary parameter to assess. Various methods for determining sequence specificity of drug interaction with natural DNA are available (reviewed in [59–62]). Of particular interest are those techniques that can detect damage to DNA from drug-treated cells, not just to naked DNA. In that way, the possible modulation of drug binding sites by nucleoprotein structures and binding competition with the entire genome can be assessed. Methods, which are based on premature termination of primer extension opposite to an adduct/lesion on the template strand, such as polymerase inhibition, ligase-mediated PCR, repetitive primer extension (RPE), or transcriptional footprinting, have been used [31,33,60,62–64]. For example, RPE allowed for comparison on the same sequences of drug effects on naked simian virus 40 (SV40) DNA, and SV40 DNA in virus-infected, drug-treated cells [24,33].

Binding motif distribution   After drug sites are experimentally mapped in selected regions, much more extensive stretches of the genome can be examined using computerized (*in silico*) sequence analysis to identify sites where DNA sequence matches preferred drug binding motif. Whereas published examples of such long-range distribution of drug binding motifs are still limited, the utility of this approach has been clearly demonstrated for several classes of drugs [24–27,33]. A computational analysis of individual, arbitrarily selected sequences can be readily performed with any software that enables restriction site analysis and customization of the analyzed motifs [25–27]. A long-range analysis, which enables batch processing of multiple GenBank entries, may require custom software [26,27], but it provides an even more powerful and versatile tool. An analysis of a large representative sample (e.g. $>10^7$ bp of human DNA sequences allows one to quickly assess where the lesions by an existing or hypothetical drug are likely to be localized [24,26,27]. The reliable predictions of the actual levels of lesions in specific regions based on the distribution of drug binding motif(s) validate the usefulness of the *in silico* analysis [24–27].

Actual lesions in specific regions   The experimental verification of drug effects in individual regions can be accomplished using various methods [5,32,65,66], including a versatile quantitative PCR (QPCR) stop assay [23,25–27,67,68]. A rigorous quantitation of region-specific lesions with these methods is essential but labor-intensive, which severely limits the number of regions that can be analyzed. However, the analysis of 1–2 arbitrarily selected regions is insufficient to assess drug region-specific potential. Fortunately, a simple computational analysis of drug binding motif distribution can guide the selection of regions that are respectively rich or poor in potential drug binding motifs [25–27]. In addition, damage to specific regions can be compared to average damage in bulk DNA, quantitated by separate methods [24,25,27]. With the advent of microarray technology and real-time PCR, various high throughput approaches to analyze drug damage in hundreds of regions should become possible in the near future eliminating the limitations of current methods.

2. Region-specific properties of cyclopropylpyrroloindole (CPI) anticancer drugs

A. Bizelesin and its distinctive damage to cellular DNA

The CPI analog bizelesin is an atypical DNA-reactive antitumor drug. With both naked and intracellular DNA, bizelesin was shown to produce both interstrand crosslinks and monoadducts [29,30,33,65]. However, interstrand crosslinks prevail and monoadducts are only a minor lesion type. The levels of crosslinks and total adducts quantitated in bulk cellular DNA are similar. For example, 6.0±0.9 crosslinks and 8.2±0.6 total adducts per 10^6 bp were observed at 100 nM bizelesin [33,65].

Unlike other clinical DNA-reactive drugs, CPI drugs, such as bizelesin and adozelesin, generate neither DNA-protein crosslinks nor strand breaks [33,65,69,70]. Also, CPI drugs do not bind RNA and proteins [71], whereas the majority of other nominally DNA-reactive drugs react extensively with cellular macromolecules. The levels of protein- and RNA-adducts formed by such drugs as nitrogen mustards typically exceed DNA adducts by 10–20 times [72,73]. In that context, CPI drugs represent agents of distinctly focused mechanism of action.

B. Specificity of bizelesin and adozelesin adducts at the nucleotide level

AT preference is a prominent feature of a CPI drug. Various studies with bizelesin-treated short model naked DNAs as well with DNA from drug-treated intact cells established the high preference of bizelesin for AT-runs with the general consensus $T(A/T)_4A$ [33,65,74,75]. A recent detailed RPE analysis of intracellularly-treated SV40 DNA demonstrated that for all the analyzed sites with the "classical" $T(A/T)_4A$ sequence, adducts were formed on both strands at a distance expected for $T(A/T)_4A$ crosslinks [33]. Furthermore, these sites amounted to a majority of all the positions implicated to be crosslinked based on analysis of drug adducts on both strands. The direct mapping of bizelesin adducts by thermal conversion of adducts to strand breaks using an SV40 MAR DNA fragment confirmed that the 5′ T-$(A/T)_4$-A 3′ motifs are the most characteristic bizelesin sites [65].

Some of the detected bizelesin binding sites reflect $A(A/T)_4A$ monoadducts [33]. Monoadducts, however, form with faster kinetics [76]. At high drug/DNA ratios used in RPE assay, rapidly formed monoadducts may thus block other drug molecules from binding to a nearby site suitable for a crosslink. At low drug/DNA ratios, however, some drug molecules are able to translocate from kinetically favored monoalkylation sites to thermodynamically favored crosslinking sites in partially overlapping nearby locations [76]. Thus, a majority of bizelesin adducts in AT-rich sites formed under pharmacologically relevant conditions of very low drug/DNA ratios is likely to correspond to interstrand crosslinks, consistent with the experimental quantitation of crosslinks and total DNA adducts [33].

Whereas the "classical" $T(A/T)_4A$ and $A(A/T)_4A$ sequences are the major target motifs for bizelesin crosslinks and monoadducts, respectively, several non-classical binding motifs were also detected. These sites are far less common than the primary motifs and seem to reflect the ability of bizelesin to form occasional non-A adducts and 7 bp-crosslinks [29,33,64,75,77]. These infrequent sites, however, are also located in generally AT-rich tracts.

Adozelesin resembles bizelesin in its sequence preferences but is markedly less selective. Adozelesin occupies 5 bp sites preferably consistent with the motif 5'-(A/T)$_{3-4}$A-3' and forms only monoadducts with DNA. Since the drug tolerates occasionally a G or C near the 5' end of the binding site [31,49,64], the minimal consensus binding sequence of adozelesin is 5'-(A/T)$_2$A-3'. Similar binding sites have been identified for the parent CPI drug, CC-1065 [31,32]. Like adozelesin, CC-1065 forms essentially monoadducts with DNA. Interestingly, however, CC-1065, but not adozelesin, increased the renaturable fraction of DNA from drug-treated cells, suggestive of interstrand crosslinks following a metabolic formation of a second alkylating site under cellular conditions [78].

The binding motifs of bizelesin and adozelesin, identified in DNA from drug-treated cells [31,33,64], should well represent actual binding preferences in the entire cellular DNA. Although these experiments analyzed only selected model sequences, (e.g. in intracellularly-treated SV40 DNA [33]), CPI drugs reacted with the entire genomic DNA. The sites in the analyzed regions thus had to compete for drug molecules with the most reactive and highest affinity sites in the entire cellular genome. Furthermore, both bizelesin and adozelesin adducted essentially identical sites in naked DNA compared to the same sequences in DNA from drug-treated cells [31,33,64]. Thus, the primary DNA sequence is the decisive factor in determining CPI drug binding, even in intact cells where DNA is organized into nucleoprotein structures.

C. Non-random lesions by CPI drugs in specific domains of SV40 DNA

Whereas the high binding specificity of CPI drugs at the nucleotide level has long been recognized, their potential for region-specific damage has been demonstrated only recently. The first experimental data suggesting that CPI drugs may form lesions with preference for certain regions came from the studies in which indirect end-labeling of thermally-induced strand breaks was used to map the distribution of drug adducts along the sequence of SV40 DNA [32,65]. Within the resolution of these determinations of approximately 50–100 bp, the parent CPI drug, CC-1065, tended to form more lesions in certain portions of the SV40 genome. Numerous adducted sites corresponding to the 5' (A/T)$_{2-3}$-A 3' motif were identified in the regions of particular vulnerability to CC-1065 [32].

As mentioned, bizelesin is more stringent than CC-1065, with preferred binding motifs being a subset of CC-1065 motifs. Accordingly, mapping of bizelesin lesions in the SV40 genome at 50–100 bp resolution showed damage to similar areas of SV40 DNA as that induced by CC-1065, although bizelesin lesions were definitely more focused (Fig. 2) [65]. The areas of the most intense bizelesin damage mapped around positions 4200, 3900, 4700 corresponding to SV40 MAR domains and ~5200, near the SV40 origin of replication (Fig. 2). Treatment of either SV40-infected cells or naked SV40 DNA resulted in a similar pattern of drug-induced lesions indicating that cellular environment does not markedly affect lesion distribution among various regions [65]. The observed patterns of bizelesin adduction at 50–100 nucleotide resolution generally followed lesion distribution predicted based on the preferred crosslinking motif 5' T-(A/T)$_4$-A 3' (Fig. 2 [65]).

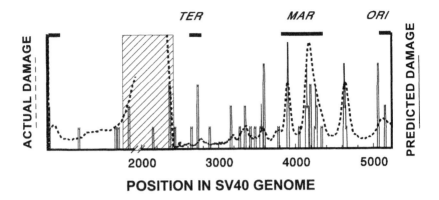

Fig. 2. Non-random damage in SV40 DNA from SV40 virus infected bizelesin-treated BSC-1 cells. Broken line shows the densitometric profile for the distribution of actual bizelesin-induced lesions measured by indirect end-labeling (except for the range of the positions between 600 and 2300, hatched block, which could not be mapped by this approach). The solid lines show the hit probabilities based on the distribution of the preferred bizelesin motif, 5'T(A/T)$_4$A3'. Adapted with permission from reference [65]. MAR, ORI, and TER indicate SV40 MAR region, origin of replication and terminus, respectively.

Bizelesin-induced interstrand crosslinks in the vulnerable regions of the SV40 genome, consistent with the 5' T-(A/T)$_4$-A 3' motif, were directly demonstrated by alkaline electrophoresis of drug-treated SV40 fragment (MAR) [65] and by adduct formation on both strands of SV40 MAR detected by the RPE technique [33].

D. Region specificity of bizelesin-induced DNA damage in cellular DNA

The affinity of bizelesin for defined SV40 regions, including replication-related domains such as SV40 MAR, suggested the possibility that non-random adduct localization in cellular DNA could be a factor in the high cytotoxic potency of bizelesin. Thus, region-specific effects of bizelesin have been investigated in human genomic DNA [25,27]. These studies also specifically addressed the possibility that bizelesin can target genomic MAR regions.

Region-specific damage by bizelesin was examined by quantitative PCR stop assay using a judiciously selected range of genomic loci [25,27]. The analyzed regions varied in their AT content, included known MAR and ORI loci, as well as regions of no known function in replication, and were located in an oncogenic locus (*c-myc*) and in non-oncogenic loci. Region selection was guided by the analysis of drug binding motif distribution. Figure 3 depicts examples of bizelesin motif distribution in two selected genes, apolipoprotein B gene (ApoB) and β-globin with indications of domains in these genes for which region-specific lesions were analyzed by QPCR stop assay. The examples of experimentally determined region-specific lesions discussed below are for bizelesin treatment of human leukemic CEM cells [25,27].

Bizelesin's preferred motif 5' T-(A/T)$_4$-A 3' is, on average, relatively infrequent. For

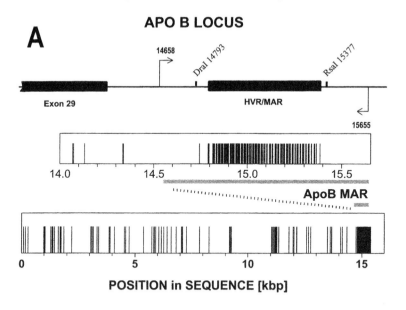

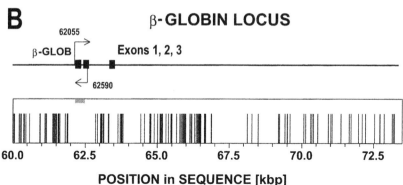

Fig. 3. Potential bizelesin binding sites in apolipoprotein B (ApoB) and β-globin genes. Panel A: the partial map of the 3′ end of ApoB gene (top) and the distribution of $T(A/T)_4A$ bizelesin binding motif (bottom). HVR/MAR indicates the hypervariable domain that coincides with ApoB MAR. Panel B: the partial map of β-globin gene (top) and the distribution of $T(A/T)_4A$ as potential bizelesin binding motifs (bottom). Regions that were amplified in QPCR stop assay experiments shown in Fig. 4 are indicated by arrows and horizontal bars. Adapted with permission from reference [25].

instance, the first 14.6 kbp of apolipoprotein B gene (ApoB, Fig. 3 top) has 123 bize-lesin sites. For comparison, a median of 33 sites is found in the same sequence for various 4 bp-recognizing restriction nucleases, with the maximal value of 120 sites. Thus, in certain stretches of DNA, bizelesin sites are similarly infrequent as sites for some restriction nucleases. However, a remarkable property of bizelesin's motif is its non-random distribution [25,27]. In some regions, such as ApoB MAR (Fig. 3 top) and, to a lesser extent, in *c-myc* MAR, clusters of bizelesin binding sites were found. Thus, the last 1 kbp of ApoB gene with the MAR region has more bizelesin sites (155, Fig. 3) than the first 14.6 kbp of the gene. In contrast, extended regions contain very few if any potential bizelesin binding sites (e.g. positions 9.3–11.0 kbp in the ApoB gene, and positions 62.0–62.6 kbp in β-globin gene, Fig. 3).

The experimental determinations by QPCR stop assay unequivocally confirm bizelesin's ability to differentially damage various regions of genomic DNA. Examples of QPCR stop assay data for ApoB MAR and the mentioned motif-poor region in the β-globin gene demonstrate profound differences in the inhibition of amplification signal (Fig. 4A). The degree of amplification inhibition is proportional to lesion frequencies in individual regions (Fig. 4B). Thus, in ApoB MAR, bizelesin-induced damage is detectable at 0.01 μM drug and reaches profound levels of 1.16 lesions/kbp at 0.1 μM bizelesin (Fig. 4B). In striking contrast, the region in the β-globin gene remains only marginally affected (0.034 lesions/kbp) at 0.1 μM bizelesin (Fig. 4B). Since this region lacks the preferred bizelesin binding sites, lesions detected at very high drug levels (1–5 μM, Fig. 4B) probably reflect adducts formed at sub-optimal sites.

How are lesions in individual regions related to average bizelesin-induced damage in the genome of CEM cells? Lesions in bulk DNA can be clearly monitored and quantitated for 0.05–0.4 μM drug [25]. The results demonstrate that absolute frequency of average lesions in bulk DNA is similar to the low lesions in the bizelesin motif-poor

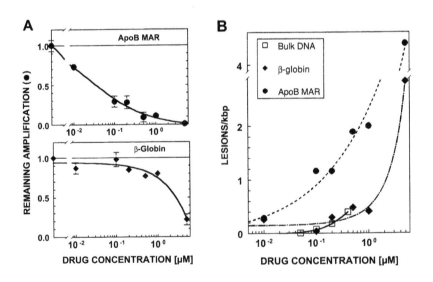

Fig. 4. Bizelesin-induced region-specific damage in DNA from drug-treated human leukemia CEM cells. Panel A: QPCR stop assay with apolipoprotein B MAR (top) and a non-MAR region in the β-globin gene. Panel B: comparison of region-specific lesions to lesions in bulk DNA. Quantitation of bizelesin lesions in specific regions is based on the inhibition of amplification in QPCR stop assay (cf. in panel A). Lesion frequencies in genomic DNA were measured after thermal conversion of adducts to breaks by sedimentation analysis in alkaline sucrose gradients. Adapted with permission from reference [25].

Table 1

Long-range *in silico* sequence analysis for the distribution of possible drug binding sites for bizelesin, adozelesin, tallimustine, U-78779, and cisplatin. For comparison, data from analogous analysis for Alu I sites as an example of restriction nuclease are shown[a]

Drug	Motif searched	Average hits/250 bp (±SD)	Maximal hits/250 bp in "hot" loci	Ratio "hot" loci to average	Expected region specificity
Bizelesin	$T(A/T)_4A$	2.8±3.6	40–99	14–34	Yes, AT islands
Adozelesin	$(A/T)_3A$	28.5±16.4	170–250	6–9	Marginal, AT islands
Cisplatin	GG, GNG, GC	50.5±13.7	97–154	2–3	No
Tallimustine	TTTTGPu	0.3±0.6	3–5	10–17	No
U-78779 Mixed A/T-G/C	$(T/A)(T/A_xG/C_y)A$[b]	22.1±11.4	90–108	4–5	No
U-78779 Pure AT	$(T/A)_6A$	6.8±7.2	112–232	16–34	Yes
Alu I	AGCT	0.7±0.2			

[a] Human sequences covering $20–43 \times 10^6$ bp were analyzed for the distribution of the indicated binding motifs. The "hits" recorded are exact matches to these motifs on both strands and given as average values per 250 bp sequence sections ("bins", cf. Fig. 5 & 7). Data are from references [23,24,26,27] except for the unpublished data for Alu I.

[b] All the permutations with y=1 or 2 vicinal G/C pairs in the central 5 bp section ($x + y = 5$) of the 7 bp binding site.

region of the β-globin gene (Fig. 4). Thus, the insensitivity of the β-globin region and other regions poor in bizelesin-preferred binding motif is probably typical of the majority of cellular DNA. In this perspective, the demonstrated vulnerability of islands of AT-rich DNA with the clusters of $T(A/T)_4A$ bizelesin sites (such as ApoB MAR) clearly documents that bizelesin is not only sequence specific but also region specific in intact cells.

E. Long-range non-random distribution of bizelesin binding motifs

Non-random distribution of potential drug binding sites can be a major property of bizelesin-induced DNA damage. To further characterize this aspect, a long range *in silico* analysis of the distribution of bizelesin binding motifs was carried out [24,27]. GenBank entries analyzed included mainly sequences of 20–300 kbp to ensure that both non-coding and coding sequences were adequately represented. The search outcomes are integrated as the number of occurrences of perfect matches to drug binding motif in 0.25 kbp segments along the analyzed sequences.

This analysis confirmed that the "density" of $T(A/T)_4A$ motif was low in the majority of human sequences (Table 1). On average, 2.8 hits/0.25 kbp (or ~11 hits/kbp) were found. However, numerous loci were identified which contained clusters of potential bizelesin binding sites with local motif density exceeding 40 hits/0.25 kbp. These loci co-localized with AT islands typically spanning 200–1000 bp and consisting of A/T repeats of various consensus motifs, occasionally interspersed with a few G/C pairs [27]. On average, the "hot" loci (with peak bizelesin motif density of 40–99 hits/0.25 kbp) occurred

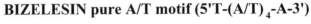

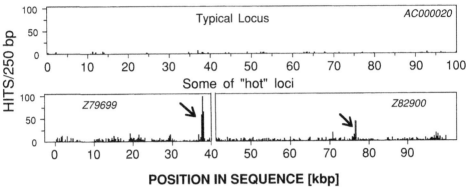

Fig. 5. Long-range distribution of potential bizelesin binding sites in human sequences (*in silico* analysis). Examples of a typical locus (GenBank AC00020, top) and hot spots for bizelesin binding (GenBank Z79699 and Z82900, bottom) from ~43×10⁶ bp examined. The histograms depict the number of perfect matches of bize- lesin preferred binding motif 5′T(A/T)₄A3′ in bins of 250 bp along the indicated GenBank sequences. Adapted with permission from reference [27].

approximately every 10^6 bp in the total of ~43 × 10^6 bp (i.e. ~1.5% of the human genome) analyzed. The histograms in Figure 5 illustrate the low levels of potential bizelesin binding sites throughout a typical sequence contrasted by profound spikes of potential binding sites in some of the hot-spot domains. Importantly, various infrequent and scattered short A/T tracts outside of the long AT islands, such as TATA boxes, are probably insignificant as potential bizelesin targets [27].

The levels of actual lesions in specific regions of DNA from drug-treated CEM cells verify the predictions of motif distribution analysis [24,25,27] (Fig. 6). ApoB MAR, a locus found to be most vulnerable to bizelesin (7.1 actual lesions/kbp/μM), also scored high in the search. Reciprocally, the highest-scoring locus among the searched 43 × 10^6 bp (GenBank Z79699, positions ~37–38 k, Fig. 5) was confirmed to be highly affected by bizelesin with ~5.5 actual lesions/kbp/μM (detected in its shorter variant AF385609, that prevails in CEM cells, Fig. 6). Direct mapping of bizelesin adducts in naked DNA of AF385609 demonstrated profound adduction sites within the ~400 bp of its 100% AT core, in contrast to only infrequent weak sites in the flanking sequences [27]. Loci that showed only low levels of DNA lesions (e.g. <0.4 lesions/kbp/μM in a β-globin locus), similar to average damage in bulk DNA, were negative also in terms of long-range motif distribution. Overall, an excellent correlation ($r^2 = 0.86$) between regional motif densities from the *in silico* analysis and actual quantitation of regional lesions (Fig. 6) demonstrate that region specificity of bizelesin parallels the presence of compatible target domains with multiple potential drug binding sites [25,27].

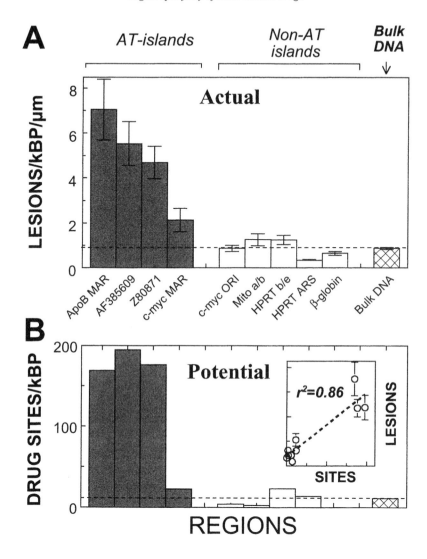

Fig. 6. Bizelesin preferentially damages MAR-like AT-rich islands over other regions and bulk DNA. Top panel: lesion frequencies in specific regions (based on QPCR stop assay systems) and in bulk DNA (by sedimentation analysis) using DNA from bizelesin-treated CEM cells. An AT island designated AF385609 is a shorter variant of the AT island in GenBank Z79699 (see Fig. 5), which is the main form present in CEM cells. Bottom panel: the density of bizelesin binding motif $T(A/T)_4A$ in individual regions and bulk DNA. "Mito a/b" denotes a region in mitochondrial DNA, whereas other regions are in nuclear DNA. Reprinted with permission from reference [27].

A. Bizelesin: preferential targeting of AT islands with clusters of drug preferred binding motifs

As discussed in detail in the previous sections, region-specific DNA damage by bizelesin has two important components: the high sequence specificity for a 6-bp motif $T(A/T)_4A$ and the existence of targets in the human genome that are compatible with this binding motif. The general significance of these two factors can be further illustrated by comparing bizelesin with other drugs. Thus, sequence- and region-specificity of AT-specific drugs of the CPI class are compared and contrasted to platinum drugs and tallimustine, representing respectively, marginally and highly specific agents at the nucleotide level. The similarities and differences among various drugs are illustrated by examples of the long-range distribution of their binding motifs (Fig. 7) and the motif distribution parameters summarized in Table 1.

B. Adozelesin vs. bizelesin: differences in sequence specificity as a predictive factor for region specific DNA damage potential

Comparison of bizelesin with adozelesin, a closely related compound with similar binding chemistry corroborates the link between sequence specificity and region specificity [25]. The relatively small differences in drug binding motifs at the nucleotide level between bizelesin and adozelesin can profoundly affect their region specificity.

Whereas adozelesin occupies only 1 bp less than bizelesin (5 vs. 6 bp, respectively) and still prefers pure A/T sites, adozelesin can tolerate G/C at the 5' portions of its binding sites. Thus, with the consensus motif $5'(A/T)_{2-4}A3'$, adozelesin has more potential binding sites in the human genome. For instance, the $5'(A/T)_3A3'$ adozelesin motif is approximately 10-fold more frequent than the bizelesin motif (Table 1). Examples of long-range motif distribution (Fig. 7) underscore the omnipresence of adozelesin motif as opposed to more localized bizelesin motif. Consequently, adozelesin forms more lesions than bizelesin in various regions [25]. For example, at 0.5 µM drug concentration, adozelesin forms 1.3 and 2 lesions/kbp in the *c-myc* ORI and *c-myc* MAR regions, respectively, compared to 0.6 and 0.7 bizelesin lesions/kbp in the same regions, respectively. In two other regions in the hypoxantine phosphoribosyltransferase (HPRT) locus, adozelesin formed 1.8 and 1.2 lesions/kbp, whereas bizelesin forms only 0.2 and 0.6 lesions/kbp, respectively.

Although adozelesin is likely to retain a marginal preference for AT islands, as illustrated by its greater binding to *c-myc* MAR over *c-myc* ORI and its motif distribution in Figure 7, adozelesin binding to other regions can be substantial. Accordingly, preliminary results show that adozelesin, but not bizelesin, damages telomeric DNA (GGGTTA)$_n$ (detected as thermally-induced strand breaks using a telomere-specific probe by Southern blotting, Napier and Woynarowski, unpublished data).

C. Platinum drugs: abundant non-region-specific lesions due to low sequence specificity

Cisplatin is known to exhibit a low sequence specificity reacting mainly with N7 guanines in the GG or GNG motifs (forming intrastrand crosslinks) and GC motifs

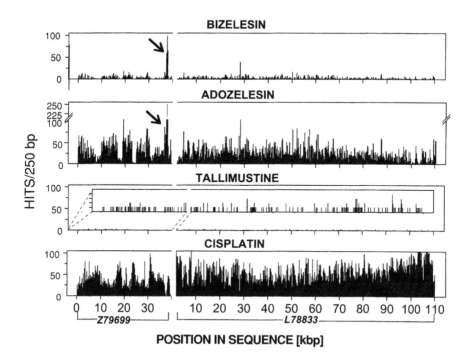

Fig. 7. Comparison of long-range distribution of potential binding sites for bizelesin, adozelesin, tallimustine and cisplatin (*in silico* analysis). The histograms depict the number of perfect matches of drug binding motifs in bins of 250 bp along GenBank sequences Z79699 and L78833. The motifs analyzed for specific drugs are listed in Table 1. Due to the very low level of tallimustine sites, tallimustine data are re-plotted with a scale expanded to a maximum at 5 hits/bin (inset). Bizelesin data for Z79699 locus with the hot-spot AT island (arrow) are re-plotted from Fig. 5. GenBank L78833 represents a typical locus for all the drugs shown, apart from a weak cluster of potential bizelesin sites ~28kbp. Adapted with permission from reference [24].

(forming interstrand crosslinks) [49,63,79–81]. These preferences are shared by various other platinum drugs [23,82].

Cisplatin binding motifs are nearly 20-fold more abundant in the genome than bizelesin motifs (Table 1). Obviously, the density of these motifs reaches peaks in GC islands, but they are ubiquitous throughout the genome. Long-range distribution of binding motifs for cisplatin underscores that virtually every section of DNA a few hundred bp long, except for pure AT islands, is likely to be damaged by cisplatin (Table 1, Fig. 7). Thus, the potential of platinum drugs for region specificity seems negligible.

Accordingly, platinum drugs show a significant damage to a wide variety of regions of genomic DNA [23,68,81,83–89] and the levels of regional damage were often found close to the levels of the overall damage measured in bulk DNA [81,86,89]. Interestingly, however, the levels of cisplatin lesions in individual regions are not uniform and seem to follow the relative density of drug binding motifs in the regions examined (except for the underprivileged mitochondrial DNA [23]). More platinum lesions are formed in GC-rich regions that contain large number of drug binding motifs. At 200 μM cisplatin, 5.5 and 5.4 lesions/kbp were found in two regions relatively rich in platinum drug binding motifs (in the β-globin and *c-myc* genes, respectively) as opposed to 1.7 lesions/kbp in a motif-poor region (in the HPRT gene) [23]. Thus, binding motif distribution reliably predicts relative vulnerability of various regions to drug adduction, which underscores the contribution of the genome factor even for agents devoid of definite region specificity. Still, due to its marginal specificity at the nucleotide level, cisplatin is too promiscuous to be classified as region specific.

D. *Tallimustine: infrequent non-region-specific lesions resulting from highly sequence-specific targeting of mixed A/T-G/C motifs*

Tallimustine, a hybrid of distamycin and a nitrogen mustard (Fig. 1), represents a drug on the opposite end of the spectrum of DNA binding preferences than cisplatin. In contrast to the nearly indiscriminate reactivity of cisplatin with N7 atoms of G moieties through the major groove, tallimustine alkylates the N3 atoms of purine residues in the minor groove. Whereas the adenine alkylation is mediated by the nitrogen mustard moiety, the distamycin portion confers the recognition of AT tracts within a definite sequence context [24,90,91]. Although the nitrogen mustard moiety has two potential alkylation sites, monoadducts have been shown to be the prevailing, if not the only, type of lesions induced by tallimustine in genomic DNA [24,90].

Studies with naked DNA confirmed tallimustine binding to AT rich tracts such as 5′-TTTTGA and 5′-TTTTAA [90,92–95]. Tallimustine adducts in intracellular DNA are also highly AT-specific with a strong preference for mixed A/T-G/C sequences such as 5′-TTTTGPu-3′ [24]. Of the four strongest drug-adduct sites identified in intracellular SV40 DNA by repetitive primer extension, three conformed to the consensus sequence of 5′-TTTTGPu-3′. Tallimustine was also found to form adducts with some other mixed T-G/C sites and in some but not all 5′-TTTTAA-3′ sequences [24].

The tallimustine consensus motif 5′-TTTTGPu-3′ features only one degenerate position in 6 bp, compared to four degenerate positions in bizelesin motif. Thus, in an absolute sense, tallimustine is more sequence specific at the nucleotide level than bizelesin. Consistent with the most stringent sequence, the tallimustine motif is approximately 10 and 170 times less frequent than motifs for bizelesin and cisplatin, respectively (Table 1) [24,27]. However, in the total of $>2 \times 10^7$ bp analyzed, the maximal number of tallimustine motifs did not exceed 4–5 hits/ 0.25 kbp section, close to the high hits in average sequence analyzed ($2.9\pm0.9/0.25$ kbp). As illustrated by examples in Figure 7, the infrequent tallimustine sites are distributed more or less uniformly throughout the genome.

This nearly random distribution of tallimustine binding motif explains the experimentally determined lack of region specificity [24]. Tallimustine does not differentiate between several discrete AT-rich regions of genomic DNA examined by QPCR stop assay. For example, 25 μM tallimustine produced an estimated level of 0.23–0.39 lesions/kbp in the specific regions, similar to lesions in bulk DNA (0.30 lesions/kbp). It needs to be re-stated that both tallimustine and bizelesin target AT-rich motifs and these motifs are distinct but not entirely dissimilar. Despite these similarities, however, tallimustine has marginal potential for region-specific damage. This paradoxical conclusion underscores the disparity between the high specificity of tallimustine sequence recognition and the lack of regions in the human genome that are significantly enriched in tallimustine binding motif.

E. *U-78779: possible region specificity but only due to the minor (pure A/T) binding motifs*

U-78779 is a bizelesin analog designed to accommodate G/C in the middle of its binding sites by inserting a pyrrole moiety between two CPI moieties [34]. Like bizelesin, U-78779 forms interstrand crosslinks. In contrast to bizelesin, however, U-78779 prefers runs of A/T interspersed with two or one G/C pairs over pure A/T motifs in naked DNA [34] and in intact cells [26], although ~1/5 of the identified sites are still in pure A/T tracts.

The *in silico* analysis indicated that the main mixed A/T-G/C motifs of U-78779 are scattered throughout the genome (Table 1) [26]. No "hot" loci were found, in which these motifs would exceed their average frequency by more than 4–5-fold (Table 1). Given the unremarkable near-uniform distribution of these A/T-G/C motifs, drug binding to these sites cannot be region-specific. In this regard, U-78779 would resemble tallimustine as a non-region specific drug, except that the collective A/T-G/C motifs of U-78779 are more abundant (22.1±11.4/0.25 kbp).

However, in contrast to the major motifs, the pure A/T motifs, also targeted by U-78779 as minor sites (5′-(A/T)$_6$A-3′, Table 1), are non-randomly distributed and cluster in the same hot loci as the pure A/T motif of bizelesin [26]. Therefore, in spite of the preference for mixed A/T-G/C at the *nucleotide level*, U-78779 is expected, like bizelesin, to show a *region* preference for AT islands, which contain clusters of (A/T)$_6$A motifs. For instance, the AT island in GenBank Z79699, targeted by bizelesin (cf., Figs 6–8), shows a very high density of such pure AT potential binding sites for U-78779.

The *in silico* predictions are fully confirmed by the experimental determinations of U-78779 lesions [26]. U-78779 adducts were found clustered in the 100% A/T core of a model AT island analyzed as naked DNA. Moreover, in drug-treated CEM cells, like bizelesin, U-78779 induces several-fold more damage to the model AT island than to a non-AT island region, consistent with the decisive contribution of pure AT binding sites. Thus, the case of U-78779 underscores the significance of the genome factor by showing that it can take precedence over drug binding preferences at the nucleotide level.

4. Biological consequences of region-specific and non-specific region damage

Given bizelesin's antitumor activity *in vivo* [96] and promising properties in clinical trials [36], the drug's ability to region-specifically target AT islands may be associated with therapeutic benefits. However, region specificity, as that found for bizelesin and U-78779, is an unprecedented property of DNA-reactive anticancer drugs. Therefore, the relevance of AT island targeting needs to be carefully scrutinized. Are region specific drugs more lethal? Why and how can targeting of AT islands lead to tumor cell growth inhibition? Studies reviewed in the subsequent sections point to AT islands as a new molecular target but also illustrate the need to better understand their nature.

A. Lethality of region-specific bizelesin lesions in comparison to non region-specific DNA damage by other drugs

Bizelesin and adozelesin are among the most cytotoxic compounds ever discovered inhibiting growth of cancer cells in sub-pico and picomolar range, respectively [24,25, 69,70] (Table 2). This extreme cytotoxicity probably results solely from extremely lethal DNA adducts, as CPI drugs do not react with other cellular macromolecules. Extrapolation of the determined frequency of bizelesin adducts in bulk DNA of CEM cells suggests that bizelesin needs to form perhaps less than 10 adducts per cell for potent antiproliferative effect (Table 2) [24,25].

Similar conclusions have been reached from studies using other cell lines and different biological endpoints. For example, bizelesin needed to form only ~100 lesions/cell for 90% inhibition of clonogenic survival of BSC-1 cells. The less sequence-specific CPI analogs, adozelesin, CC-1065, remain prominently cytotoxic, although they may need to generate more DNA lesions than bizelesin for an equitoxic effect [25,32,69,97,98].

AT-specific lesions can thus result in a potent antiproliferative effect, although AT specificity alone does not guarantee the utmost lesion lethality. Region-specific lesions induced by bizelesin remain unmatched and seem to be as much as two orders of magnitude more lethal than AT-specific, but not region-specific, tallimustine lesions (Table 2). Lesions by U-78779, some of which are probably in AT islands, seem to be similarly lethal as bizelesin lesions [26]. These observations are compatible with the notion that hits in long AT islands can be more critical to cell survival than scattered hits in various AT-tracts, such as those formed by tallimustine. An alternative explanation could be that interstrand crosslinks, a major form of bizelesin lesion [33], represent an inherently more lethal type of DNA damage than lesions formed by tallimustine, i.e. monoadducts [24].

Table 2
Cytotoxicity and DNA adducts of region-specific and non region-specific drugs – lethality of DNA lesions induced by bizelesin, tallimustine, and cisplatin

Cell line	Drug	Cytotoxicity GI_{50} [M] [b]	Region-specific lesions	Adducts in bulk DNA/ kbp/μM	Estimated lesions/cell at GI_{50} [c]
CEM (tumor)[a]	Bizelesin	0.6×10^{-12}	Yes	0.88	2
	Tallimustine	3.5×10^{-9}	No	0.022	220
	Cisplatin	1.5×10^{-6}	No	0.032	14000
WI-38 (normal)[d]	Bizelesin	$(\sim 8 \times 10^{-11})$	Yes	0.95	~200

[a] Data from references [23–25].

[b] Drug concentration inhibiting net cell growth by 50% based on MTT assay or cell counts (in parenthesis).

[c] Estimated by linear extrapolation of lesion frequencies /kbp/μM to the GI_{50} concentration assuming 2.9×10^9 bp/cell (rounded values).

[d] Reference [117].

This possibility, however, seems less likely, since it is inconsistent with the profound lethality of adozelesin-induced monoadducts and dramatically lower lethality of lesions by cisplatin, which comprise an appreciable proportion of interstrand crosslinks [99,100]. Other factors beyond lesion localization, for instance different repair patterns may conceivably influence lesion lethality. However, the lethality of AT-specific lesions by bizelesin, adozelesin, U-78779, and tallimustine seems to follow their relative potential to target AT islands. This trend lends further support to the idea that region-specific targeting of AT islands leads to more grave consequences than uniformly distributed DNA lesions.

Although non region-specific tallimustine lesions are less lethal than those by CPI

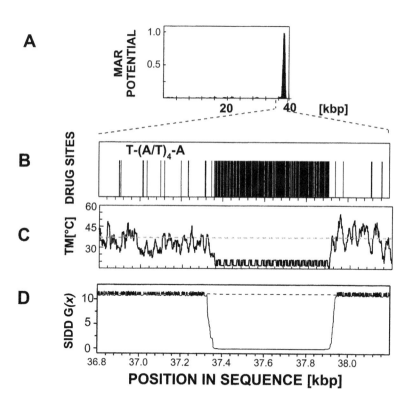

Fig. 8. MAR potential of AT islands targeted by bizelesin and the duplex instability. Examples of sequence attributes for AT island in GenBank Z79699. Panel A: calculated MAR potential for GenBank Z79699. Middle Panel B: clusters of bizelesin binding sites in AT-island in Z79699 sequence that exactly coincides with the peak in MAR potential (cf. also Fig. 5 for long-range distribution of bizelesin binding motif in these sequences). Panel C: thermodynamic duplex stability at each nucleotide position in Z79699 sequence calculated as melting temperatures of a 15 bp oligonucleotide "moving window". Horizontal broken line in the duplex stability panel corresponds to an average duplex stability across the entire sequence. Panel D: superhelical stress-induced duplex destabilization calculated in Z79699 sequence. Data adapted from reference [27].

drugs, they still seem to be quite infrequent at cell growth inhibitory concentrations (Table 2) [24]. Tallimustine is also markedly more cytotoxic than simple nitrogen mustards [39]. Interestingly, tallimustine-targeted sequences are intrinsically bent, and tallimustine adduction of $(A)_n$-tracts is expected to cause their un-bending. It is possible that the distortion of local DNA bending plays a role in cellular effects of tallimustine [24].

By contrast to lesions in AT-tracts, GC-specific cisplatin adducts seem to be markedly less cytotoxic (~50-fold) than tallimustine–DNA adducts and dramatically less lethal (by several orders of magnitude) than bizelesin lesions formed preferentially in AT islands (Table 2) [24]. The actual lethality of cisplatin lesions may be even lower than suggested by data in Table 2, given that non-DNA lesions, which may also contribute to cytotoxic effects [100,101], exceed many times the number of DNA adducts. To the author's knowledge, none of the other alkylating drugs that prefer simple G/C motifs approaches cytotoxicity levels (and probably also lesion lethality) comparable to those of bizelesin.

B. MAR function of AT islands targeted by bizelesin

What is the nature of bizelesin-targeted AT islands? Can these domains constitute a critical target in the functional sense? Based on several lines of evidence, we proposed [27] that these AT islands represent domains with a high potential to function as matrix associated regions, MARs, i.e. sites critical for defining the borders of DNA loops and organizing their attachment to the nuclear matrix.

Several of the identified AT islands are known MARs, for example, *c-myc* MAR, ApoB MAR, and SV40 MAR [102–104]. Recently, the Z79699 AT island, the "hottest" identified region for bizelesin binding motif, has been found to bind specifically to isolated nuclear matrices (Woynarowski *et al.*, unpublished data). The distribution of potential bizelesin binding sites is matched closely by domains identified as strong MAR candidates [27] using the MAR-Finder tool [105]. The peaks of computed "MAR potential" (for example see Fig. 8A) are in exactly the same location as the clusters of bizelesin binding sites (Fig. 8B).

One particular property, common to AT-rich MAR domains, is the profound destabilization of DNA duplexes resulting in stress-induced base unpairing [106–108]. Bizelesin-targeted long AT islands coincide with thermodynamically destabilized regions (Fig. 8C) that are likely to be unwound under the physiological levels of supercoiling (Fig. 8D) [27]. The partially base-unpaired native conformation of such AT-rich elements is thought to be related to topological separation of DNA loops by MAR domains. Elimination of such base-unpaired sites from MAR sequences by mutations abrogates their function [108]. Likewise, over-stabilization of DNA duplexes by bizelesin-induced interstrand crosslinks, which prevents strand separation, is expected to eliminate the native partially unwound conformation of AT islands [33].

Furthermore, bizelesin crosslinks can distort (straighten) bent elements [77]. A bent AT-tract in the SV40 ORI region is essential for replicon initiation, and bend elimination by mutation inhibits ORI-dependent replication [109]. Similar bent AT-tracts are also typically present in cellular MAR domains [52,108,110]. Thus, bizelesin binding to AT-rich MAR-like regions is probably associated with various profound structural changes that may interfere with their MAR functions. Given the fundamental role of MAR domains in the organization and functioning of nuclear chromatin, bizelesin-induced distortions are likely to have profound effects on cell function.

In particular, the high lethality of lesions in MAR regions can be related to the critical role of MAR domains in replication [58,111]. Bizelesin is a potent and specific inhibitor of genomic DNA replication, with orders of magnitude lower effect on RNA synthesis and no detectable effect on protein synthesis [112]. Other CPI drugs with the potential to affect AT islands invariably showed extremely potent antireplicative effects [32,113–115]. Analysis of nascent genomic DNA synthesized after bizelesin treatment demonstrated that bizelesin inhibits first new replicon initiation [112]. The effect on chain elongation required higher drug levels and prolonged incubation. Inhibition of replicon initiation by bizelesin was also observed in a replication model with defined replicon size [112]. Intriguingly, only a single bizelesin adduct per several replicons seems sufficient for a nearly complete inhibition of replicon initiation [112]. The inhibition at the level of replication control is consistent with the postulated role of prominent MAR domains in the nearly simultaneous firing of each replicon in replicon clusters that constitute replication factories [53,116]. Possibly, bizelesin-induced distortion of the atypical MAR duplex structures can propagate, impeding the function of an entire replication factory.

C. Hypersensitivity to AT island-targeting DNA damage

Whereas bizelesin is generally highly cytotoxic, the observed differences in the sensitivity of several tumor and normal cell lines span several orders of magnitude [117]. Some tumor cell lines appear to be significantly ($p<0.05$) hypersensitive relative to other tumor and normal cell lines examined. Similar trends are seen for cytotoxic potency of U-78779, which also is region specific for AT islands. In contrast, adozelesin, which binds various short AT tracts in addition to AT islands, does not produce such bimodal responses. For human leukemia CEM and normal WI-38 cells, we have additionally demonstrated that the differential sensitivity to bizelesin is not due to any detectable differences in drug reactivity with cellular DNA, or lesion repair. In both cell lines, bizelesin produced similar levels of lesions in bulk DNA (0.88 and 0.95 lesions/kbp/μM, respectively), which were essentially irreversible upon the postincubation in drug-free medium. Also, bizelesin maintains its regional preference for AT islands in normal WI-38 cells. However, bizelesin is markedly less cytotoxic against WI-38 cells ($GI_{50}=\sim80$ pM) than against CEM cells ($GI_{50}=0.2$–0.6 pM, Table 2). Thus bizelesin adducts are ~2 orders of magnitude more lethal in tumor CEM cells than in normal WI-38 cells (Table 2). The bimodal sensitivity to AT island damage suggests the existence of fundamental molecular differences among various cancer and normal cells, possibly at the genomic level.

These differential responses to bizelesin might be related to the minisatellite nature of AT islands. Mini- and microsatellites are inherently unstable, hypervariable loci and their changes are likely to follow or even contribute to the tumorigenic phenotype [118,119]. Minisatellite instability can result in extensive repeat length polymorphism and genomic restructuring, e.g. through induction of fragile sites [120]. In the case of bizelesin-targeted AT islands, their instability might lead to their amplification and/or reorganization in tumor cells, including acquiring MAR function [117]. Clearly, further studies are needed to elucidate in depth the nature of the differences in the organization of MAR-like AT islands and the consequences of the interference with such regions in various tumor and normal cells.

5. Conclusions: paradigm for region specificity of small molecular weight agents

Although the reviewed studies have been centered on AT-specific drugs, the systematically established determinants of region-specific damage to cellular DNA should be generally applicable to small molecular weight drugs. The following paragraphs discuss several aspects where such generalization can be made.

Sequence specificity is necessary but insufficient   As exemplified by cisplatin, drugs with low sequence specificity, which recognize 2–3 bp, cannot be region specific. Recognition of longer motifs (such as 6 bp for bizelesin) may be necessary for region specificity. Even seemingly small differences in binding motifs, such as those between bizelesin and adozelesin, can result in poor region specificity of the more promiscuously binding drug. However, high specificity at the nucleotide level per se may not lead to significant region specificity, if drug lesions are scattered throughout the genome nearly randomly (like those induced by tallimustine). Thus, various highly sequence-specific drugs may, paradoxically, share the lack of regional specificity with non sequence-specific agents, such as cisplatin.

Targets with clusters of drug binding sites. The genome factor   As proposed in individual studies [24–27], a universal paradigm for small molecular weight drugs can be that: *region specificity requires sequence specificity matched to a genomic target with clusters of potential drug binding sites*. A drug's ability to affect a critical region at multiple sites (clusters of drug binding motif) can be more important than just sequence specificity at the nucleotide level. Bizelesin is region specific because its potential binding sites are well matched to the targeted regions in the human genome. The notion of matching drug binding motif to a genomic target is applicable also to drugs with multiple binding motifs, as exemplified by the case of U-78779. The majority of lesions by U-78779 are unlikely to be region specific, since the main drug binding motifs are distributed nearly randomly. Yet, U-78779 is expected to target the same AT islands as bizelesin, but due to its minor, pure AT motifs. Thus, the genome factor can take precedence over drug binding preferences at the nucleotide level.

Targeting non-coding repetitive sequences   Controlled manipulation of gene expression is one potential use of region-specific strategies that has attracted considerable attention. However, targeting repetitive sequences, which tend to localize in non-coding, yet often crucial, regions, is a relevant alternative. Non-coding repetitive sequences are more suitable as targets for small molecular weight drugs than unique sites in the coding regions. Domains of repetitive sequences can provide clusters of drug binding sites and constitute a relatively abundant target, in contrast to unique coding sequences. Such abundant repetitive targets could be selectively hit even with agents of less-than-ideal specificity.

MAR-like AT island minisatellites provide a proof of principle that targeting repetitive sequences crucial for the survival of cancer cells is feasible. Other attractive targeting possibilities certainly exist. For instance, highly abundant telomeric DNA provides a potentially critical target. Conceivably, telomeric repeats $(GGGTTA)_n$ could be precisely matched by "modular" drugs that would at the same time recognize $(A/T)_2A$ portion (like adozelesin) and GG portion, like cisplatin.

Identification of new targets and optimization of drug binding properties The results with several classes of drugs suggest a practical utility and benefits of considering region specificity as a component of the drug development process. In particular, long-range *in silico* analysis for potential drug binding sites can reliably predict whether a specific binding motif is likely to result in preferential regional damage. For an existing drug with known antitumor activity, this approach may help to uncover new, potentially important targets, such as bizelesin-targeted MAR-like AT islands. Conversely, for a known critical target, one can readily establish which binding motifs would maximize a preference for the desired targeted domain over non-target loci. Such information can guide the design of actual drugs with the desired (and already optimized) targeting properties. Importantly, in addition to anticancer strategies, analogous approaches can explore the genomes of various pathogenic microorganisms with the idea of targeting repetitive DNA domains that are absent from the human genome.

6. Summary

A non-random distribution of DNA lesions in various domains of the human genome may be important for the selectivity of antiproliferative action of anticancer drugs. Recent investigations explored the determinants of region-specific damage to cellular DNA in cancer cells by comparing highly AT-specific drugs (bizelesin, adozelesin, U-78779, and tallimustine) to a relatively non-specific drug, cisplatin. These studies consistently show that sequence specificity at the nucleotide level and a long-range distribution of drug binding motifs analyzed "*in silico*" predict well the actual relative vulnerability of specific regions to various drugs. Bizelesin is the first small molecular weight anticancer drug demonstrated to preferentially damage specific regions in cellular DNA. Bizelesin is region specific because its binding motif is non-randomly distributed, with clusters in long (200–1000 bp) islands of AT-rich DNA. A substantial specificity of this drug at the nucleotide level is an important and necessary factor to accomplish region-specific damage. The less specific adozelesin targets similar regions as bizelesin but with only a marginal preference over other areas of the genome. Cisplatin is not a region-specific drug, due to its low specificity of DNA recognition and the omnipresence of its potential binding sites. High specificity at the nucleotide level, however, is per se insufficient for region specificity. Tallimustine is not region specific, despite the very high sequence specificity. Paradoxically, the highly specific, and accordingly infrequent, tallimustine sites are distributed throughout the genome nearly randomly, resembling the distribution of much less specific (and more numerous) lesions by cisplatin. The consequences of DNA lesions in AT-islands targeted by bizelesin seem to be dramatically more lethal than non region-specific lesions by tallimustine and cisplatin. The reviewed findings provide a proof of principle that targeting of critical repetitive sequences (not necessarily coding regions), which allow for clustering of drug binding motif, can be the paradigm for region specificity of small molecular weight agents. The notion of regional drug targeting should be recognized as an essential extension of the rational design of sequence-specific drugs.

Acknowledgements

I am grateful to all members of my laboratory, in particular Dr Maryanne Herzig, for their competent and creative contributions to the reviewed investigations. This research has been supported in part by a grant CA71969 from the National Cancer Institute and Pilot Project grants from Children's Cancer Research Center at the University of Texas Health Science Center at San Antonio. I would like also to acknowledge a partial support from the Department of Radiation Oncology, University of Texas Health Science Center at San Antonio and the sustained encouragement from its chair, Dr Terence Herman.

References

[1]   K.W. Kohn, J.A. Hartley, W.B. Mattes, *Biochem. Pharmacol.* 37 (1988) 1799–1800.
[2]   J.A. Hartley, J.W. Lown, W.B. Mattes, K.W. Kohn, *Acta Oncol.* 27 (1988) 503–510.
[3]   P.B. Dervan, *Science* 232 (1986) 464–471.
[4]   P.B. Dervan, *Bioorg. Med. Chem.* 9 (2001) 2215–2235.
[5]   D.A. Gewirtz, M.S. Orr, F.A. Fornari, J.K. Randolph, J.C. Yalowich, M.K. Ritke, L.F. Povirk, R.T. Bunch, *Cancer Res.* 53 (1993) 3547–3554.
[6]   R.T. Bunch, L.F. Povirk, M.S. Orr, J.K. Randolph, F.A. Fornari, D.A. Gewirtz, *Biochem. Pharmacol.* 47 (1994) 317–329.
[7]   C. Giovannangeli, C. Helene, *Antisense Nucleic Acid Drug Dev.* 7 (1997) 413–421.
[8]   D.M. Gowers, K.R. Fox, *Nucleic Acids Res.* 27 (1999) 1569–1577.
[9]   K.R. Fox, *Curr. Med. Chem.* 7 (2000) 17–37.
[10] G.E. Plum, D.S. Pilch, S.F. Singleton, K.J. Breslauer, *Ann. Rev. Biophys. Biomol. Struct.* 24 (1995) 319–350.
[11] L.J. Maher, 3rd, *Cancer Invest.* 14 (1996), 66–82.
[12] D.E. Wemmer, P.B. Dervan, *Curr. Opin. Struct. Biol.* 7 (1997) 355–361.
[13] S. Neidle, *Biopolymers* 44 (1997) 105–121.
[14] B.S. Reddy, S.K. Sharma, J.W. Lown, *Curr. Med. Chem.* 8 (2001) 475–508.
[15] P.G. Baraldi, B. Cacciari, A. Guiotto, R. Romagnoli, A.N. Zaid, G. Spalluto, *Curr. Pharm. Des.* 4 (1998) 249–276.
[16] J.B. Chaires, *Curr. Opin. Struct. Biol.* 8 (1998) 314–320.
[17] S. White, E.E. Baird, P.B. Dervan, *Chem. Biol.* 4 (1997) 569–578.
[18] S. White, E.E. Baird, P.B. Dervan, *Biochemistry* 35 (1996) 12532–12537.
[19] J. Cho, M.E. Parks, P.B. Dervan, *Proc. Natl. Acad. Sci. USA* 92 (1995) 10389–10392.
[20] N.R. Wurtz, P.B. Dervan, *Chem. Biol.* 7 (2000) 153–161.
[21] S. Neidle, M.S. Puvvada, D.E. Thurston, *Eur. J. Cancer* 4 (1994) 567–568.
[22] W.B. Mattes, J.A. Hartley, K.W. Kohn, D.W. Matheson, *Carcinogenesis* 9 (1988) 2065–2072.
[23] J.M. Woynarowski, W.G. Chapman, C. Napier, M. Herzig, P. Juniewicz, *Mol. Pharmacol.* 54 (1998) 770–777.
[24] M.C. Herzig, A.V. Trevino, B. Arnett, J.M. Woynarowski, *Biochemistry* 38 (1999) 14045–14055.
[25] J.M. Woynarowski, C. Napier, A.V. Trevino, B. Arnett, *Biochemistry* 39 (2000) 9917–9927.

[26] M.C.S. Herzig, K. Rodriguez, A. Trevino, B. Arnett, J. Dziegielewski, L. Hurley J.M. Woynarowski; *Biochemistry*, 41 (2002) 1545–1555.

[27] J.M. Woynarowski, A.V. Trevino, A.V.K. Rodriguez, S.C. Hardies, C.J. Benham, *J. Biol. Chem.* 276 (2001) 40555–40566.

[28] M. Warpehoski, *Drugs of the Future* 16 (1991) 131–141.

[29] C.S. Lee, N.W. Gibson, *Biochemistry* 32 (1993) 2592–2600.

[30] D. Sun, H.J. Park, L.H. Hurley, *Chem. Res. Toxicol.* 6 (1993) 889–894.

[31] K.L. Weiland, T.P. Dooley, *Biochemistry* (1991) 7559–7565.

[32] M.M. McHugh, J.M. Woynarowski, M.A. Mitchell, L.S. Gawron, K.L. Weiland, T.A. Beerman, *Biochemistry* 33 (1994) 9158–9168.

[33] J.M. Woynarowski, W.G. Chapman, C. Napier, M.C.S. Herzig, *Biochim. Biophys. Acta.* 1444 (1999) 201–217.

[34] H.J. Park, R.C. Kelly, L.H. Hurley, *J. Am. Chem. Soc.* 118 (1996) 10041–10051.

[35] M. Cristofanilli, W.J. Bryan, L.L. Miller, A.Y. Chang, W.J. Gradishar, D.W. Kufe, G.N. Hortobagyi, *Anticancer Drugs* 9 (1998) 779–782.

[36] G.H. Schwartz, C. Aylesworth, J. Stephenson, T. Johnson, E. Campbell, L. Hammond, D.D. Von Hoff, E.K. Rowinsky, *Proc. Amer. Soc. Clin. Oncol.* 19 (2000) 235a.

[37] D.L. Boger, T.V. Hughes, M.P. Hedrick, *J. Org. Chem.* 66 (2001) 2207–2216.

[38] P.G. Baraldi, G. Balboni, M.G. Pavani, G. Spalluto, M.A. Tabrizi, E. De Clercq, J. Balzarini, T. Bando, H. Sugiyama, R. Romagnoli, *J. Med. Chem.* 44 (2001) 2536–2543.

[39] F.M. Arcamone, F. Animati, B. Barbieri, E. Configliacchi, R, D.A., C. Geroni, F.C. Giuliani, E. Lazzari, M. Menozzi, N. Mongelli, *J. Med. Chem.* 32 (1989) 774–778.

[40] G. Pezzoni, M. Grandi, G. Biasoli, L. Capolongo, D. Ballinari, F.C. Giuliani, B. Barbieri, A. Pastori, E. Pesenti, N. Mongelli, *et al.*, *Brit. J. Cancer* 64 (1991) 1047–1050.

[41] C.J. Punt, Y. Humblet, E. Roca, L.Y. Dirix, R. Wainstein, A. Polli, I. Corradino, *Brit. J. Cancer* 73 (1996) 803–804.

[42] J. Viallet, D. Stewart, F. Shepherd, J. Ayoub, Y. Cormier, N. DiPietro, W. Steward, *Lung Cancer* 15 (1996) 367–373.

[43] W.L. Walker, M.L. Kopka, D.S. Goodsell, *Biopolymers* 44 (1997) 323–334.

[44] C. Bailly, J.B. Chaires, *Bioconjug. Chem.* 9 (1998) 513–538.

[45] Y. Sugiura, T. Matsumoto, *Biochemistry* 32 (1993) 5548–5553.

[46] M.M. McHugh, J.M. Woynarowski, L.S. Gawron, T. Otani, T.A. Beerman, *Biochemistry* 34 (1995) 1805–1814.

[47] R.J. Cobuzzi Jr., S.K. Kotsopoulos, T. Otani, T.A. Beerman, *Biochemistry* 34 (1995) 583–592.

[48] J.A. Hartley, J.P. Bingham, R.L. Souhami, *Nucleic Acids Res.* 20 (1992) 3175–3178.

[49] G.J. Bubley, G.K. Ogata, N.P. Dupuis, B.A. Teicher, *Cancer Res.* 54 (1994) 6325–6329.

[50] J.B. Smaill, J.Y. Fan, P.V. Papa, C.J. O'Connor, W.A. Denny, *Anticancer Drug Des.* 13 (1998) 221–242.

[51] P.N. Cockerill, W.T. Garrard, *Cell* 44 (1986) 273–282.

[52] T. Boulikas, *J. Cell. Biochem.* 52 (1993) 14–22.

[53] R. Berezney, M.J. Mortillaro, H. Ma, X. Wei, J. Samarabandu, *Int. Rev. Cytol.* 162A (1995) 1–65.

[54] S.M. Stack, L.K. Anderson, *Chromosome Res.* 9 (2001) 175–198.

[55] M. Ramakrishnan, W.M. Liu, P.A. DiCroce, A. Posner, J. Zheng, T. Kohwi-Shigematsu, T.G. Krontiris, *Mol. Cell. Biol.* 20 (2000) 868–877.

[56] S.V. Razin, K. Shen, E. Ioudinkova, K. Scherrer, *J. Cell. Biochem.* 74 (1999) 38–49.

[57] E. Bonnefoy, M.T. Bandu, J. Doly, *Mol. Cell Biol.* 19 (1999) 2803–2816.

[58] F. Wanka, *Acta Biochim. Pol.* 42 (1995) 127–131.

[59] J.A. Hartley, M.D. Wyatt, *Methods Mol. Biol.* 90 (1997) 147–156.

[60] K.A. Grimaldi, J.A. Hartley, *Methods Mol. Biol.* 90 (1997) 157–180.

[61] K.A. Grimaldi, J.P. Bingham, J.A. Hartley, *Methods Mol. Biol.* 113 (1999) 227–240.

[62] K.A. Grimaldi, S.R. McAdam, J.A. Hartley, *Methods Mol. Biol.* 113 (1999) 241–255.

[63] C. Cullinane, G. Wickham, W.D. McFadyen, W.A. Denny, B.D. Palmer, D.R. Phillips, *Nucleic Acids Res.* 21 (1993) 393–400.

[64] C.S. Lee, G.P. Pfeifer, N.W. Gibson, *Biochemistry* 33 (1994) 6024–6030.

[65] J.M. Woynarowski, M. McHugh, L.S. Gawron, T.A. Beerman, *Biochemistry* 34 (1995) 13042–13050.

[66] K. Wassermann, K.W. Kohn, V.A. Bohr, *J. Biol. Chem.* 265 (1990) 13906–13913.

[67] D.P. Kalinowski, S. Illenye, B. Van Houten, *Nucleic Acids Res.* 20 (1992) 3485–3494.

[68] K.A. Grimaldi, J.P. Bingham, R.L. Souhami, J.A. Hartley, *Anal. Biochem.* 222 (1994) 236–242.

[69] T.J. Zsido, T.A. Beerman, R.L. Meegan, J.M. Woynarowski, R.M. Baker, *Biochem. Pharmacol.* 43 (1992) 1817–1822.

[70] C.S. Lee, N.W. Gibson, *Cancer Res.* 51 (1991) 6586–6591.

[71] V.L. Reynolds, J.P. McGovren, L.H. Hurley, *J. Antibiot.* 39 (1986) 319–334.

[72] E.W. Vogel, M.J. Nivard, *Mutat. Res.* 305 (1994) 13–32.

[73] P.D. Lawley, D.H. Phillips, *Mutat. Res.* 355 (1996) 13–40.

[74] M. Lee, M.C. Roldan, M.K. Haskell, S.R. McAdam, J.A. Hartley, *J. Med. Chem.* 37 (1994) 1208–1213.

[75] D. Sun, L.H. Hurley, *J. Am. Chem. Soc.* 115 (1993) 5925–5933.

[76] S.J. Lee, F.C. Seaman, D. Sun, H.P. Xiong, R.C. Kelly, L.H. Hurley, *J. Am. Chem. Soc.* 119 (1997) 3434–3442.

[77] A.S. Thompson, L.H. Hurley, *J. Mol. Biol.* 252 (1995) 86–101.

[78] A. Skladanowski, M. Koba, J. Konopa, *Biochem. Pharmacol.* 61 (2001) 67–72.

[79] K.A. Grimaldi, S.R. McAdam, R.L. Souhami, J.A. Hartley, *Nucleic Acids Res.* 22 (1994) 2311–2317.

[80] K. Hemminki, W.G. Thilly, *Mutat. Res.* 202 (1988) 133–138.

[81] M.M. Jennerwein, A. Eastman, *Nucleic Acids Res.* 19 (1991) 6209–6214.

[82] V. Murray, J. Whittaker, W.D. McFadyen, *Chem Biol. Interact.* 110 (1998) 27–37.

[83] G.C. Wang, L.M. Hallberg, E.W. Englander, *Mutat. Res. DNA Repair* 434 (1999) 67–74.

[84] S.S. Daoud, M.K. Clements, C.L. Small, *Anti-Cancer Drugs* 6 (1995) 405–412.

[85] D.P. Kalinowski, S. Illenye, B. Van Houten, *Nucleic Acids Res.* 20 (1992) 3485–3494.

[86] F. Oshita, A. Eastman, *Oncology Res.* 5 (1993) 111–118.

[87] F. Oshita, H. Arioka, Y. Heike, J. Shiraishi, N. Saijo, *Jap. J. Cancer Res.* 86 (1995) 233–238.

[88] F. Oshita, N. Saijo, *Jap. J. Cancer Res.* 85 (1994) 669–673.

[89] Z. Zhang, M.C. Poirier, *Chem. Res. Toxicol.* 10 (1997) 971–977.

[90] M. Broggini, H.M. Coley, N. Mongelli, E. Pesenti, M.D. Wyatt, J.A. Hartley, M. D'Incalci, *Nucleic Acids Res.* 23 (1995) 81–87.

[91] M.D. Wyatt, M. Lee, B.J. Garbiras, R.L. Souhami, J.A. Hartley, *Biochemistry* 34 (1995) 13034–13041.

[92] M. Broggini, M. Ponti, S. Ottolenghi, M. D'Incalci, N. Mongelli, R. Mantovani, *Nucleic Acids Res.* 17 (1989) 1051–1059.

[93] M. Fontana, M. Lestingi, C. Mondello, A. Braghetti, A. Montecucco, G. Ciarrocchi, *Anti-Cancer Drug Des.* 7 (1992) 131–141.

[94] M.D. Wyatt, B.J. Garbiras, M.K. Haskell, M. Lee, R.L. Souhami, J.A. Hartley, *Anti-Cancer Drug Des.* 9 (1994) 511–525.

[95] P. Beccaglia, K.A. Grimaldi, J.A. Hartley, S. Marchini, M. Broggini, M. D'Incalci, *Brit. J. Cancer.* 73 (1996) 12.

[96] D.L. Walker, J.M. Reid, M.M. Ames, *Cancer Chemother. Pharmacol.* 34 (1994) 317–322.

[97] T.J. Zsido, J.M. Woynarowski, R.M. Baker, L.S. Gawron, T.A. Beerman, *Biochemistry* 30 (1991) 3733–3738.

[98] R.D. Hightower, B.U. Sevin, J. Perras, H. Nguyen, R. Angioli, M. Untch, H. Averette, *Cancer Invest.* 11 (1993) 276–282.

[99] A.C. Plooy, A.M. Fichtinger-Schepman, H.H. Schutte, M. van Dijk, P.H. Lohman, *Carcinogenesis* 6 (1985) 561–566.

[100] J.M. Woynarowski, S. Faivre, M.C.S. Herzig, B. Arnett, W.G. Chapman, A.V. Trevino, E. Raymond, S.G. Chaney, A. Vaisman, M. Varchenko, P.E. Juniewicz, *Mol. Pharmacol.* 58 (2000) 920–927.

[101] J.M. Woynarowski, S. Koester, B. Woynarowska, B. Arnett, A.V. Trevino, D. Chan, A. Higdon, R. Munoz, M.C.S. Herzig, S. Faivre, *Proc. Amer. Assoc. Cancer. Res.* 40 (1999) 736.

[102] R.H. Chou, J.R. Churchill, M.M. Flubacher, D.E. Mapstone, J. Jones, *Cancer Res.* 50 (1990) 3199–3206.

[103] E. Boerwinkle, W.J. Xiong, E. Fourest, L. Chan, *Proc. Natl. Acad. Sci. USA* 86 (1989) 212–216.

[104] Y. Pommier, P.N. Cockerill, K.W. Kohn, W.T. Garrard, *J. Virol.* 64 (1990) 419–423.

[105] J.A. Kramer, G.B. Singh, S.A. Krawetz, *Genomics* 33 (1996) 305–308.

[106] C. Benham, T. Kohwi-Shigematsu, J. Bode, *J. Mol. Biol.* 274 (1997) 181–196.

[107] W.A. Krajewski, S.V. Razin, *Mol. Gen. Genet.* 235 (1992) 381–388.

[108] J. Bode, Y. Kohwi, L. Dickinson, T. Joh, D. Klehr, C. Mielke, T. Kohwi-Shigematsu, *Science* 255 (1992) 195–197.

[109] J.A. Borowiec, *J. Virol.* 66 (1992) 5248–5255.

[110] R. Berezney, *J. Cell Biochem.* 47 (1991) 109–123.

[111] P.A. Dijkwel, J.L. Hamlin, *Int. Rev. Cytol.* 162A (1995) 455–484.

[112] J.M. Woynarowski, T.A. Beerman, *Biochim. Biophys. Acta* 1353 (1997) 50–60.

[113] L.H. Li, D.H. Swenson, S.L. Schpok, S.L. Kuentzel, B.D. Dayton, W.C. Krueger, *Cancer Res.* 42 (1982) 999–1004.

[114] R.J. Cobuzzi Jr., W.C. Burhans, T.A. Beerman, *J. Biol. Chem.* 271 (1996) 19852–19859.

[115] M.M. McHugh, S.R. Kuo, M.H. Walsh-O'Beirne, J.S. Liu, T. Melendy, T.A. Beerman, *Biochemistry* 38 (1999) 11508–11515.

[116] R. Berezney, D.D. Dubey, J.A. Huberman, *Chromosoma* 108 (2000) 471–484.

[117] J. Jackson, A.V. Trevino, M.C.S. Herzig, J.M. Woynarowski, *Proc. Amer. Assoc. Cancer Res.* 43 (2002) 1092.

[118] D.I. Smith, H. Huang, L. Wang, *Int. J. Oncol.* 12 (1998) 187–196.

[119] P.R. Turner, W.A. Denny, *Mutat. Res. Fundam. Mol. Mech. Mutagen.* 355 (1996) 141–169.

[120] G.R. Sutherland, E. Baker, R.I. Richards, *Trends Genet.* 14 (1998) 501–506.

ELSEVIER

Advances in DNA Sequence-specific Agents 04
(2002) 29–46

Advances
in DNA
Sequence-specific
Agents

www.elsevier.nl/locate/series/adna

DNA-alkylating events associated with nitrogen mustard based anticancer drugs and the metabolic byproduct Acrolein

Michael E. Colvin[1]*, Judy N. Quong[2]

[1]*Biology and Biotechnology Research Program, Computational and Systems Biology Division, Mailstop L-448, Lawrence Livermore National Laboratory, Livermore CA 94550, and*
[2]*Chemistry and Materials Science Directorate, Mailstop L-231 Lawrence Livermore National Laboratory, Livermore CA 94550*

Abstract

The nitrogen mustard anticancer drugs are widely used for treating many forms of cancer. However, a number of toxic side effects, including secondary tumors, limit their clinical utility. There is strong evidence that the effectiveness for killing cancer cells (cytotoxicity) of these drugs is directly related to the ability of their metabolites to form interstrand DNA crosslinks. These crosslinks are formed by the sequential alkylation of two bases on complementary DNA strands. Other products, including monoalkylated adducts and intrastrand crosslinks are much less cyto-toxic, and can mutate the DNA in normal cells leading to the formation of secondary tumors. The nitrogen mustards have a complex DNA-alkylation chemistry. They form monoalkylation adducts at a number of different sites on DNA and the predominant lesions vary between mustards and are strongly modulated by conjugation with DNA-targeting groups including intercalators, minor groove binders, and triplex-forming DNA oligomers. Only a small fraction of the DNA-alkylation events proceed to the formation of the therapeutically useful, highly cytotoxic crosslinks. For this reason there is considerable ongoing research to improve the efficiency of crosslink formation, and reduce the mutagenic effects of the other mustard-induced lesions. The phosphoramide mustard prodrugs including cyclophosphamide and ifosfamide, produce acrolein as a metabolic byproduct. Like the mustards themselves, acrolein has a complex DNA-alkylation chemistry, leading to a diversity of DNA adducts.

*Corresponding author
E-mail address: Colvin2@LLNL.GOV
Advances in DNA Sequence-specific Agents, Volume 4, 29–46

1 Introduction

The nitrogen mustard based DNA-alkylating agents were the first effective anticancer drugs and remain important drugs against many forms of cancer. More than fifty years of research on the nitrogen mustards has yielded a broad range of therapeutically useful compounds and a detailed knowledge of the biochemical mechanism of these drugs. Nevertheless, there is much ongoing research on the phosphoramidic and other nitrogen mustards to increase their potency and reduce their toxic and mutagenic side effects [1].

Since the therapeutic activity of these drugs involves DNA interstrand crosslinking, there has been extensive research on how the chemical properties of the nitrogen mustard alkylating agents and the sequence and structure of DNA affect this crosslinking. Several recent review articles describe the chemistry of DNA-crosslinking agents [2–4]. In this review, we will focus on the DNA-alkylating properties of the nitrogen mustard and phosphoramide mustard based crosslinking agents, and highlight some examples of chemical simulations applied to understanding the properties of crosslinking agents. Additionally, we will provide an overview of the DNA-alkylating properties of acrolein, an important metabolic byproduct of the phosphoramide mustard prodrugs.

2. Overview of therapeutic activity of nitrogen mustards

Chemical alkylating agents, first developed for use as chemical weapons, were later found to be efficacious in the treatment of cancer [5]. In particular, members of the phosphoramidic mustard prodrug class, cyclophosphamide and ifosfamide (Fig. 1a,b), are now widely used in cancer chemotherapy [1].

These alkylating agents, or their active metabolites, cyclize to reactive intermediates which then form covalent bonds with sites on nucleic acids that result in a variety of types of damage including depurination and base mispairing [6]. Based on observations that bifunctional, versus monofunctional, alkylating agents exhibit greater cytotoxic potency, the most therapeutically useful lesions appear to be DNA interstrand crosslinking events that can lead to strand scission, inhibition of DNA synthesis and, ultimately, cell death [7,8].

In contrast to direct-acting agents such as mechlorethamine (Fig. 1d), the phosphoramidic mustard prodrugs (Fig. 1a–c) require metabolic and chemical activation to liberate an active phosphoramidic mustard. Overwhelming evidence indicates that the nitrogen mustards and the active metabolites of the phosphoramidic mustard prodrugs alkylate DNA via a highly electrophilic cyclic aziridinium species formed from an intramolecular displacement reaction in which the nitrogen in the mustard group attacks the β carbon, displacing the chloride (Fig. 2). The therapeutically useful nitrogen and phosphoramidic mustards contain at least two chloroethyl groups, allowing for the sequential formation of two activated aziridinium species. Once activated to the highly electrophilic aziridinium form, the final step in the therapeutic activity is the crosslinking of DNA by sequential reactions typically at two guanine N7 sites of opposite strands of DNA. DNA monoadducts and intrastrand adducts have also been observed (see below).

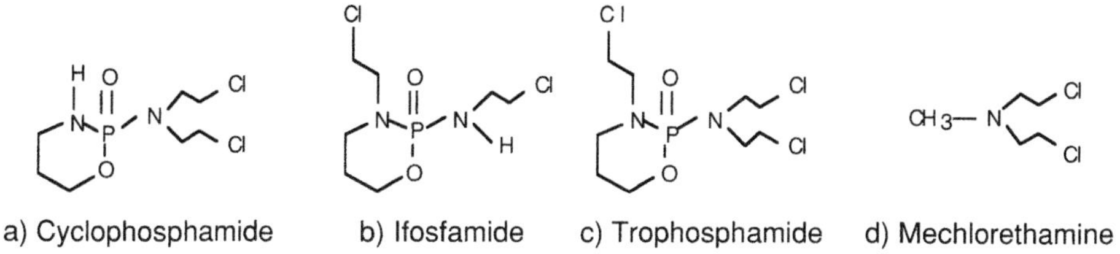

a) Cyclophosphamide   b) Ifosfamide   c) Trophosphamide   d) Mechlorethamine

Fig. 1. (a–c) Three phosphoramidic mustard-based anticancer prodrugs. These compounds require metabolic activation to a reactive species that forms DNA crosslinks between guanine bases on opposite DNA strands. (d) A nitrogen mustard anticancer drug that forms interstrand crosslinks without metabolic activation.

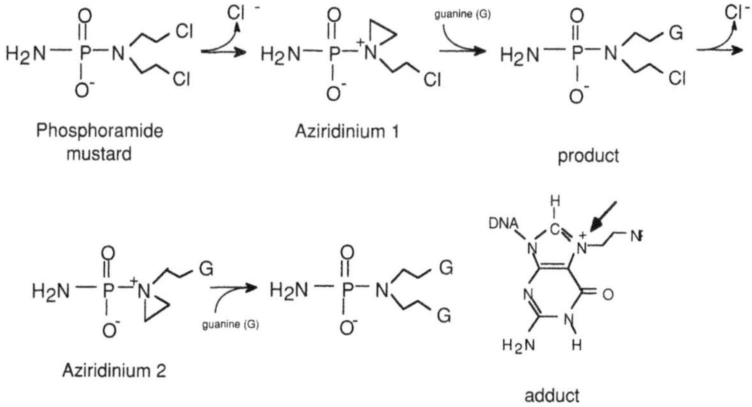

Fig. 2. Chemical reactions leading to therapeutically active DNA crosslink as shown for phosphoramide mustard, the active metabolite of cyclophosphamide. First the phosphoramide mustard undergoes intramolecular cycliza-tion to form a highly electrophilic aziridinium ion, which subsequently reacts at the guanine N7 position (or other cellular nucleophile). This process is repeated for the second alkylation reaction. Note that P–N bond hydrolysis could occur at any point in this process, but has been shown to preferentially occur for the aziridinyl intermediates.

The clinical usefulness of these drugs is diminished by their undesirable side effects including bladder, cardiac and reproductive toxicity, as well as secondary cancers [9], some of which are due to non-crosslinking DNA lesions (monoalkylated and intrastrand bisalkylated) and metabolic byproducts of the prodrugs, such as acrolein. Despite these side effects, there is an ongoing scientific interest in these agents, both because of their continuing clinical value, and because emerging data on the action of these compounds, as well as co-therapeutics, offers the possibility of developing analogs with significantly higher therapeutic indices.

3. Monofunctional adduction sites

The original observation by Brookes and Lawley that nitrogen mustards primarily adduct DNA at the guanine N7 site has been substantiated by many studies for a wide variety of nitrogen mustards [10]. More recently, several other potential nitrogen mustard adduction sites have been identified, including the adenine N3 [11] and the monophosphate [12]. When conjugated to DNA-targeting groups such as intercalators or minor groove binders, nitrogen mustards have been observed to alkylate at a number of other sites including adenine N7 and N1 [13]. A detailed survey of the DNA-alkylation sites of many antitumor agents including more than a dozen nitrogen mustards has recently been published [14].

The distribution of preferred DNA-alkylation sites is in part due to the accessibility of these sites in double stranded DNA as evidenced by the effect of DNA-targeting groups conjugated to nitrogen mustards (see below). However, there is also evidence that the nucleophilic reactivity of the different sites also plays a role. For example, Kikuchi *et al.* [15] performed semiempirical quantum chemical calculations on the reaction of dimethyl aziridinium ion with nucleophilic sites on the DNA bases. They found that for the isolated bases, the guanine O6 was the preferred site of alkylation, followed by the guanine N7, adenine N3, and adenine N7. Inclusion of the complementary cytosine which hydrogen bonds to the guanine O6 greatly increased the reaction barrier for alkylation at this site. Therefore, for the reaction by aziridinium ion on DNA helices, the predicted alkylation sites are consistent with the observed alkylation patterns. However, it should be noted that reaction barriers calculated by semiempirical quantum chemical methods frequently have very large errors, therefore revisiting this aziridinium-DNA reaction process using more accurate first principles quantum chemical methods and a solvent model is warranted.

4. Sequence specificity of nitrogen mustard alkylation

Although the nitrogen mustards are not generally selective for guanines in specific DNA sequences [9], Kohn *et al.* [16] and others [17,18] have observed that many mustards, including phosphoramide mustard, mechlorethamine, and chlorambucil preferentially bind at guanines located on the interior of guanine clusters. For example, Mattes *et al.* studied the alkylation patterns of four different nitrogen mustards on a single 3741 base pair segment of DNA that contained runs of up to five contiguous guanines in many different sequence contexts [17]. All of the alkylating agents showed a marked preference for monoalkylating at the guanine N7 position within multiple guanine runs, with the maximum alkylation occurring in runs of four guanines. Interestingly, two of the compounds, uracil mustard and quinacrine mustard, also showed enhanced alkylation at isolated guanines in certain sequence contexts. This sequence specificity for these two mustards was subsequently verified in studies of guanine alkylation patterns in intact human myeloid leukemic cells [19].

The preference for guanine alkylation within multiple guanine runs has been explained in terms of the molecular electrostatic potential of DNA, which is a measure at different points in space of the attraction or repulsion felt by an ion [20]. Kohn *et al.* found that the regions of DNA most attractive to a positive ion correlated well with the observed binding pattern for all compounds studied (except uracil mustard) [16]. They also observed that the overall most electrostatically attractive regions occurred at the center position of

GGG sequences. This suggests that the association of the mustard to DNA occurs *after* the formation of the charged aziridinium species. Their suggestion that the aziridinium ion's charge guides the binding site is supported by Hartley *et al.* [21] who showed that increased ionic strength and cationic DNA affinity binders would alter the sequence selectivity of the nitrogen mustards. An interesting consequence of this hypothesis may be that mustards carrying other charges near the aziridinium ion (such as phosphoramide mustard) that will affect their long-range electrostatic interactions, may show differential selectivity to sequence. A more detailed modeling study of the selectivity of aziridinium ion to DNA was performed by Pearlstein *et al.* [22] who used molecular mechanics to study the interaction of dimethyl aziridinium ion with B-DNA duplex tetramer. They found an energy minimum for the aziridinium ion near the N7 site of an internal guanine in the 5'-d(GGGG) sequence, but did not find a minimum near any N7 site in the 5'-d(GAGA) sequence, consistent with Kohn's results.

The electrostatic interactions invoked above to explain the increased alkylation rates within multiple guanine runs can also be exploited to design nitrogen mustard analogs with greatly enhanced alkylation rates. For example Cullis *et al.* have recently shown that conjugating linear polyamines to chlorambucil increases the guanine N7 alkylation up to 10 000-fold [23]. They used molecular modeling to demonstrate that this rate enhancement likely results from the strong binding of the cationic polyamines to the anionic DNA phosphodiester backbone that favorably positions the mustard moiety adjacent to the guanine N7 in the major groove.

Beyond the use of simple polycationic targeting groups, much research has gone into attaching DNA groove-binding [24,25], intercalating [26,27], and even triplex-forming DNA oligomers [28] to nitrogen mustard crosslinking agents (see recent review in this series [13]). A dramatic example of nitrogen mustard alkylating patterns being affected by DNA-targeting groups is the work of Hartley and co-workers on benzoic acid mustard covalently tethered to 1, 2, or 3 pyrrole-amide units [29–31]. The benzoic acid mustard and its monopyrrole conjugate produce primarily the guanine N7 monoadducts typical of the nitrogen mustards. However, the di- and tripyrrole conjugates alkylate only in poly-AT tracts, producing adenine N3 (minor groove) adducts. The tripyrrole benzoic acid mustard conjugate exhibited high sequence specificity, alkylating primarily at the 5'-TTTTGPu sequence. Furthermore, they found that this highly sequence selective tripyrrole conjugate was significantly more cytotoxic than its monopyrrole conjugate. A review of minor-groove targeting of nitrogen mustard compounds has recently been published [32].

The conjugation of nitrogen mustards to DNA intercalating agents has also been shown to increase the alkylation efficiency and change the alkylation patterns. For example Prakash *et al.* studied the effect of covalently linking aniline mustards to the DNA intercalator 9-aminoacridine using several different chemical linkers [33]. For several of the linkers tested, the aminoacridine-bound mustards showed increased levels of alkylation and increased cytotoxicity relative to other nitrogen mustards, although the fraction of monoalkylated adducts that went on to form interstrand crosslinks was not improved over the ~5% typically seen for nitrogen mustards. Another interesting result of this work was that as the aminoacridine-aniline mustard linker chain was increased in length, the favored alkylation site changed from guanine N7 to the adenine N7 (also in the major groove).

Kutyavin *et al.* studied DNA alkylation by chlorambucil covalently bound to either or both ends of a triplex-forming DNA oligomer of 20 basepairs [28]. They found that the DNA-bound chlorambucil would efficiently alkylate guanines in a target DNA double helix at sites consistent with the formation (pre-alkylation) of a DNA triple helix that orients the chlorambucil adjacent to a run of three guanines. Subsequent studies by the same group using DNA-bound chlorambucil and benzoic acid mustard demonstrated that the preferred alkylation site was guanine N7, followed by adenine N3, and that both monoalkylation products and interstrand crosslinks were formed [34].

5. Structural properties of DNA crosslinks

As described above, the therapeutic cytotoxicity of the phosphoramidic mustard prodrugs and nitrogen mustards is believed to involve interstrand DNA crosslinking; however, there are a number of other possible DNA-alkylation products. The earliest evidence that DNA crosslinking caused the cytotoxic (and anticancer) activity of the nitrogen mustard alkylating agents was the much greater potency of the bifunctional versus monofunctional (non-crosslinking) compounds [35]. Subsequent studies have found more direct evidence correlating the cytotoxicity of several nitrogen mustards with the level of interstrand crosslinking induced in cells [36]. Mechanistically, the cytotoxic effectiveness of the nitrogen mustards is well grounded in experimental data showing that interstrand crosslinks are relatively long-lived lesions [9] that inhibit both DNA replication and transcription [37], and are poorly repaired relative to DNA monoadducts [38]. Figure 3 shows the range of observed alkylation products, including (d) the monoalkylated adduct, (e) bisalkylated intrastrand alkylation, and (f–h) bisalkylated interstrand crosslinks.

The original experiments demonstrating that the mustards form DNA interstrand crosslinks were based on the analysis of fully digested DNA and therefore could not define the exact configuration of the guanine–guanine crosslink. Initially, these agents were assumed to crosslink in a 1–2 5'-CG configuration, based on the estimated N7–N7 distances in the Watson-Crick model of B-DNA [10]. The 5'-GC 1–2 crosslink configuration was shown to be more energetically feasible by Hausheer *et al.* [39] on the basis of molecular dynamics simulations of crosslinked 8 basepair DNA duplexes, including full water and counter ion solvation shells. Note that this result is not necessarily inconsistent with a 1–3 crosslinking configuration, which has been shown by subsequent simulations to be nearly isoenergetic with the 1–2 crosslink. This study also described quantum chemical calculations on the phosphoramide mustard adducted guanine that suggest the intriguing hypothesis, based on N7–N7 distances in the DNA helix, that the second alkylation reaction may occur via direct S_N2 substitution by the guanine N7 on the mustard chloroethyl group. Additionally, this study hypothesized that a thymine flanking either guanine would inhibit crosslinking.

Subsequent experimental studies have provided unambiguous evidence that crosslinks occur in a 1–3 configuration crosslinking 5'-d(GNC) for mechlorethamine [40,41], chlorambucil [42], and phosphoramide mustard [43]. The structural consequences of such a crosslink have been the subject of several studies. Rink *et al.* [44] pointed out that the N7–N7 distance in B-DNA of 8.9 Å is too long to allow the cross-link by the

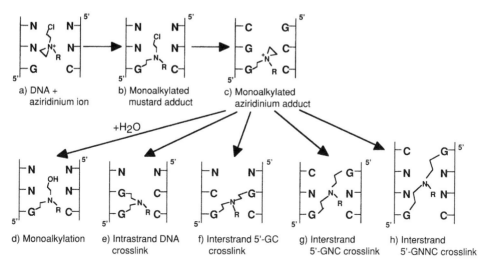

Fig. 3. Sequence and outcomes of phosphoramide or nitrogen mustard alkylation reactions with DNA. (a,b,c) First three structures in the alkylation process, starting with the reaction of DNA with the activated mustard compound; (d,e) final monoalkylated and intrastrand crosslink products that can lead to DNA mutation and carcinogenesis; (f–h) interstrand crosslink products believed responsible for the therapeutic cytotoxicity.

ethyl-amine-ethyl linker (maximum extension of ~5.1 Å) without structural perturbation. They subsequently used molecular mechanics minimizations and electrophoretic mobility measurements to show that mechlorethamine cross-linking induced a bend in DNA, although neither the orientation nor magnitude of the bend could be characterized [45].

Dong *et al.* [43] reported molecular dynamics simulations on nor-nitrogen mustard crosslinking a 12 basepair DNA duplex in both 1–2 and 1–3 interstrand conformations. They also simulated the dynamics of unadducted DNA in order to monitor the natural guanine N7 distances. These simulations were performed in spheres of approximately 2000 water molecules for 200 picoseconds. Additionally, quantum chemical calculations were performed on the nitrogen mustard crosslink at structures sampled from the molecular dynamics simulations. This study showed that the 1–2 and 1–3 crosslinks were nearly equal in energy, so that the observed preference for the 1–3 configuration cannot be explained simply in terms of greater stability. Instead, they concluded that the 1–3 crosslink is favored because that conformation is more kinetically accessible in the fluctuating DNA duplex structure. Although the 1–3 N7–N7 distance is on average longer than the 1–2 distance (9.0 Å vs. 7.9 Å), the 1–3 distance shows much greater fluctuations, occasionally moving within 0.03 Å of the crosslinked distance, compared to 0.40 Å for the closest approach in the 1–2 configuration. This result is similar to that from a molecular mechanics study of DNA crosslinked by a seven-member alkyl chain that found that the observed 5′-GNC crosslink to be isoenergetic with the 5′-GC crosslink and therefore must be kinetically, rather than thermodynamically, favored [46].

Bauer and Povirk [47] used molecular dynamics simulations of DNA monoadducted by the aziridinium intermediates of melphalan, mechlorethamine, and phosphoramide mustard to explain the relative lack of intrastrand crosslinks observed for melphalan compared to the other two mustards. They monitored the distance from the aziridinium carbons to the N7 of both the adjacent guanine, and the guanine on the opposite strand. Over their 100 picosecond simulation, they found that the average distance from the aziridinium carbons to the adjacent (intrastrand) guanine N7 was 3.14 Å for mechlorethamine and 3.36 Å for phosphoramide mustard — in both cases close enough for the alkylation reaction to occur. In contrast, the melphalan formed a stable conformation with its aziridinium carbons more than 6 Å from the intrastrand N7, because of a hydrogen bond between the melphalan ammonium group and the phosphate and 5'-oxygen two bases from the adduction site. They concluded that this precluded the formation of the intrastrand crosslink by melphalan. They did not report the distances from the aziridinium ion to the opposite N7 position, so it is not clear whether a favorable conformation for forming the interstrand crosslink was observed in their simulation. Note that these molecular dynamics simulations were performed in the absence of explicit water molecules and counter ions, but instead used a screening method to approximate the effect of aqueous solvation.

Although most of the research on nitrogen mustard crosslinking has focused on the guanine N7 to guanine N7 crosslinks, other crosslinked products have been observed and may be related to the biological activity of certain nitrogen mustards. For example, Osborne and Lawley have reported adenine N3 to guanine N7 melphalan crosslinks of high molecular weight DNA [11]. Under some reaction conditions they found that these adenine–guanine crosslinks were formed at higher frequency than the typical guanine–guanine crosslinks. Since the adenine N3 is in the DNA minor groove, and the guanine N7 in the major groove, these crosslinks must lead to considerable distortion of the DNA double helix. Osborne and Lawley suggest that these highly distorted crosslinks might be the cause of the high immunoreactivity of melphalan-induced crosslinks. Osborne *et al.* subsequently found evidence for guanine–adenine crosslinks from mechlorethamine although the yield, under the conditions studied, was considerably lower than that for guanine–guanine crosslinks [48]. Note that in neither study was it possible to conclusively rule out that the adenine–guanine crosslinks are experimental artefacts occurring after the DNA hydrolysis.

Interestingly, Romero *et al.* have recently shown that mechlorethamine will crosslink mismatched basepairs within DNA double helices [49]. In particular they found that cytosine–cytosine mismatch pairs would exhibit crosslinking rates up to 27.5% depending on the local DNA sequence. They also used molecular dynamics simulations to relate the structure of the DNA near the mismatch pair to the crosslinking rates. Although the crosslinking of mismatched basepairs has not been proposed to have any therapeutic anticancer utility, this work demonstrates that the nitrogen mustards may prove useful as probes for C–C mismatches that are characteristic of some diseases.

6. Biological consequences of monofunctional adducts

Given the evidence indicating that the cytotoxic potency of the nitrogen mustard anticancer drugs arises from the interstrand crosslinks, it is significant that only a small fraction of their DNA-alkylation products actually form such crosslinks (typically 5–10%) [9]. Even if the other, non-crosslinking DNA-alkylation products have no toxic consequences, the large doses required to counter this low crosslinking efficiency would be best avoided. This is particularly true for cyclophosphamide and its analogs for which much of the acute toxicity results from a metabolic byproduct, acrolein (see below).

The exact role of the non-crosslinked DNA adducts in the observed mutagenicity and carcinogenicity of the nitrogen mustards is a complex issue. A number of early studies indicated that DNA monofunctional adducts might be the primary cause of the nitrogen mustard mutagenicity [50]. More recently, several studies that compare the mutagenic potency of nitrogen mustards with their non-crosslinking "half-mustard" analogs, have shown that the DNA interstrand crosslinks are strongly mutagenic when inaccurately repaired by the cell, and are actually more potent mutagens than the corresponding monofunctional adducts when evaluated at equal concentrations [38,51]. However, as Ferguson and co-workers have pointed out, at doses of equal cytotoxicity, the monofunctional and bifunctional (crosslinking) mustards have nearly equal mutagenic potencies [52].

There is still much more to be learned about the complex factors causing these drugs to be both effective cytotoxic therapeutics and potent mutagens and carcinogens. Moreover, development of co-therapeutics that diminishes the mutagenic potency of the crosslinking agents without lowering their cytotoxicity provides a new dimension for optimizing the anticancer effects of these drugs. For example, O^6-benzylguanine was recently found to dramatically affect the action of nitrogen mustard crosslinking agents [53]. In particular O^6-benzylguanine reduces the mutagenicity of the crosslinking phosphoramide mustard to the level of a non-crosslinking PM analog, while increasing the cytotoxicity of the phosphoramide mustard. O^6-benzylguanine is known to be a non-toxic modulator of O^6-alkylguanine-DNA transferase (AGT) [54], however activity towards nitrogen mustards appears to be completely unrelated to depletion of AGT, and current evidence indicates that it acts by modulating the cell cycle, in particular by inducing G1 arrest.

7. Factors influencing DNA crosslinking efficiency

Information about the relationship between the structure and efficiency of crosslinking agents can be gleaned from experimental data on the type of interstrand crosslink formed by nitrogen mustards [40–42,44] and phosphoramidic mustards [43,55], as well as other crosslinking agents including α,ω,-alkanediol-dimethylsulfonate esters (CH_3–SO_2–O–$(CH_2)_n$–O–SO_2–CH_3) [56] and diepoxyalkanes [57,58]. These different crosslinking agents are shown in Figure 4. These compounds cover a range of crosslinking lengths from 6.3 Å to 11.3 Å and therefore provide insight into the relationship between these lengths and crosslinking efficiency.

Table 1 summarizes the crosslink preferences observed for these guanine N7–N7 crosslinking agents with different linker chain lengths. The units listed for each compound type are proportional to the amount of crosslink formed for each type, but are not intercomparable between types. For each crosslinking agent, the fully extended chain length is given in angstroms and number of atoms and for each DNA target the guanine N7–N7 distance is given in angstroms [56].

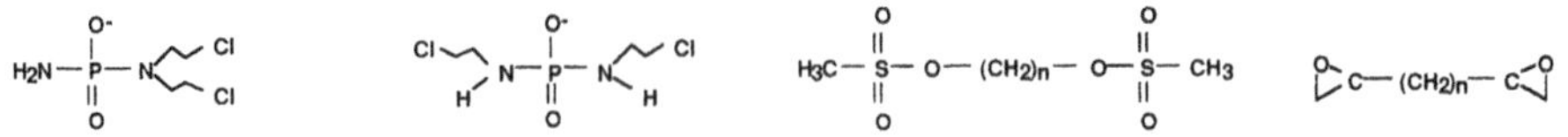

Phosphoramide Mustard (PM) (a | Isophosphoramide mustard (IPM) (b) | Sulfonate esters (c) | Diepoxyalkanes (d)

Fig. 4. Structures of DNA crosslinking agents with different linker lengths. (a,b) the active metabolites of the phosphoramidic mustard prodrugs cyclophosphamide and ifosfamide, which have crosslinking lengths of 5 and 7 atoms respectively; (c) sulfonate esters and (d) diepoxyalkanes with variable crosslink lengths.

A significant feature of this data is that in many cases the observed crosslinker lengths are considerably shorter than the N7–N7 distances in undistorted B-DNA. This discrepancy is made more significant by the fact that the transition state necessary for the second alkylation step is actually ~1 Å shorter than the final crosslink. Another key finding from Table 1 is that, regardless of the crosslinker length, all compounds that exhibit interstrand crosslinking favor the 5′-GNC crosslink (Fig. 3g). The variation in crosslinking efficiency with chain length is more complex. For the mustards, IPM, with a 10.2 Å maximal crosslinking length shows slightly higher amounts of 5′-GNC crosslink product than the other compounds, but curiously shows much less 5′-GNNC crosslinking than do the

Table 1
Comparison of observed product distributions for different crosslinking agents and different guanine N7–N7 sequences. The units listed for each compound type are proportional to the amount of crosslink formed for each type, but are not intercomparable between types. For each crosslinking agent the fully extended chain length is given in terms of angstroms and number of atoms in the crosslinking chain. For each DNA target, the guanine N7–N7 distance is given in angstroms

Bifunctional alkylating agents	Crosslinker length Å (# atoms)	5′-GC N7–N7 = 7.2 Å	5′-GNC N7–N7 = 9.0 Å	5′-GNNC N7–N7= 12.3 Å
Nitrogen & phosphoramidic mustards [55]		Relative autoradiogram band intensity		
Mechlorethamine	7.4 Å (5)	< 0.0012	0.086	0.011
Phosphoramide mustard	7.4 Å (5)	< 0.0012	0.067	0.013
Isophosphoramide mustard	10.2 Å (7)	< 0.0012	0.094	0.0045
Sulfonate ester [56]		Relative autoradiogram band intensity		
$CH_3SO_3(CH_2)_4O_3SCH_3$	6.3 Å (4)	0.0	0.0	0.0
$CH_3SO_3(CH_2)_5O_3SCH_3$	7.5 Å (5)	1.1 ± 0.1	10.7 ± 3.4	1.1 ± 0.1
$CH_3SO_3(CH_2)_6O_3SCH_3$	8.7 Å (6)	1.0 ± 0.0	29.1 ± 9.4	1.1 ± 0.1
$CH_3SO_3(CH_2)_8O_3SCH_3$	11.3 Å (8)	1.1 ± 0.1	10.7 ± 3.4	1.1 ± 0.1
Diepoxyalkanes [58]		% of original compound forming crosslink		
(CH_2OCH)–(CH_2OCH_2)	6.3 Å (4)	< 0.1%	1.0 ± 0.4%	0.9 ± 0.3%
(CH_2OCH)–$(CH_2)_2$–$(CHOCH_2)$	8.7 Å (6)	< 0.1%	0.3 ± 0.2%	0.1 ± 0.0%
(CH_2OCH)–$(CH_2)_4$–$(CHOCH_2)$	11.3 Å (8)	< 0.1%	1.1 ± 0.5%	0.3 ± 0.1%

shorter mustard compounds. Similarly, the sulfonate esters and diepoxyalkanes show a non-obvious relationship between the crosslinker length and the N7–N7 distance in the target sequence.

This complex relationship between crosslinker lengths and the preferred crosslink configuration has been attributed to two possible factors (that could act simultaneously): 1) the possibility that the monoalkylated N7 product induces a bend in the DNA that shortens the N7–N7 distance, and 2) the high flexibility of solution phase DNA that may lead to very close N7–N7 distances at reasonable frequency. The possibility that the DNA is actually distorted by the monoalkylation event has indirect experimental support from studies showing that cations, such as the aziridinium ion, bound to DNA induce bending [59], and has been studied by molecular dynamics simulations. Remias *et al.* [60] modeled chlorambucil bound to a DNA 7-mer in both non-covalent, mono-adduct, and crosslinked configurations. They concluded from their simulations that the non-covalent binding of the chlorambucil aziridinium ion near the guanine N7 orients the local DNA into a conformation favorable for crosslinking.

These experimental and simulation data clearly indicate that the length of the crosslinking arm is a significant variable in the crosslinking efficiency. The optimal crosslinker length will involve a balance between being sufficiently long to allow facile crosslinking (i.e. without requiring major distortion of the DNA), and being sufficiently short that the second alkylation reaction is statistically likely to occur with the nearby guanine N7, rather than to other cellular nucleophiles including water.

Changing the length of the chloroalkyl arms in the phosphoramidic mustard analogs will affect both the final interstrand crosslinking as well as the activation and alkylation reactions. Early studies by Arnold *et al.* [35] explored analogs of cyclophosphamide and nor-nitrogen mustard with modified chloroalkyl arms. Interestingly, they found that nor-nitrogen mustard analogs containing mixed chlorethyl and chloropropyl arms did not change the therapeutic index (DL_{50}/DC_{50}), but replacing cyclophosphamide chloroethyl groups with chloropropyl or chlorobutyl, decreased the therapeutic index. However, these early studies did not discriminate between the various factors that could lead to decreased potency, so that the specific effects on crosslinking efficiency have not been established. Hence, if these prove efficient at forming crosslinks *in vitro*, therapeutically effective crosslinking agents might still be developed from these analogs.

The effect of varying chain length on the kinetics of intramolecular cyclization is known to be complex, with relative rate constants for the 3- versus 5-membered ring formation varying from 20 000 to 0.001 [61]. Published cyclization rates for the ω-halogenoalkylamines, that are chemically analogous to the mustards, suggest that the chlorobutyl derivative of PM will undergo intramolecular cyclization more rapidly than the chloroethyl or chloropropyl forms [62]. However, once formed, the larger ring system will have considerably less ring strain than the aziridinium and therefore should be a more selective electrophile.

Structurally, the longer chloroalkyl arm will also have several effects. The monoalkyl product will have a longer crosslinking distance (e.g. approximately 1 Å per $-CH_2-$ unit), which should facilitate formation of the crosslink. Additionally, the larger rings will be more hydrophobic because of their increased alkyl content and therefore may preferentially orient within the DNA major groove.

8. DNA binding by acrolein

Acrolein arises as a byproduct of cyclophosphamide and ifosfamide metabolism [63]. This and other α,β-unsaturated carbonyl compounds also occur as combustion byprod-ucts of organic compounds and thus are found in tobacco smoke, automobile exhaust, and vapors of cooking oils [64]. Acrolein is a known mutagen and toxin [65–69] and is the simplest member of the class of electrophilic α,β-unsaturated carbonyl compounds that includes other known mutagens such as crotonaldehyde (Fig. 5). Acrolein is highly reactive and conjugates with sulfhydryl groups present in proteins and peptides, and binds to and depletes cellular nucleophiles such as glutathione [70] resulting in a variety of toxic effects including disruption of metabolism [71] and alterations in cell proliferation and membrane transport [72]. Acrolein also interacts with DNA resulting in the forma-tion of cyclic DNA adducts [73–76], frameshift interactions, strand breaks [77], and DNA crosslinking [78–80]. In this review, we focus on exocyclic DNA adducts. These adducts have been associated with the induction of secondary tumors of the bladder in cyclophos-phamide-treated patients [66] and are believed to be responsible for cyclophosphamide and acrolein mutagenicity and have been proposed as a carcinogen biomarkers in chemo-prevention trials [66,81–84]. More recently, evidence shows that acrolein DNA adducts may have a positive effect, aiding in the therapeutic effects of cyclophosphamide (albeit, also potentially increasing the risk of secondary tumors via acrolein-induced mutagen-esis) [63,70,85].

Acrolein adducts resulting from the *in vitro* reaction with each of the nucleobases have been studied and identified. In all cases, the reaction of the nucleobase and acrolein gener-ates exocyclic products in which a new ring is fused to positions on the bases that are normally involved in Watson-Crick hydrogen bonding. The exocyclic adduct follows a two-step reaction pathway involving a Michael addition followed by ring closure.

Two major 1,N2′-propanoguanosine adducts were identified after the reaction of de-oxyguanosine and acrolein under physiological conditions using a combination of HPLC with UV, mass, and NMR spectral techniques [73]. These results were verified using HPLC and electrochromatography [86]. These two adducts, 3-(2′-deoxyribosyl)-5,6,7,8-tetra-hydro-8-hydroxypyrimido[1,2-a]purine-10-one and 3-(2′-deoxyribosyl)-5,6,7,8 tetra-hydro-6-hydroxypyrimido[1,2-a]purine-10-one, differ by the position of the hydroxyl group on the purine ring (Fig. 5a). The adduct with the hydroxyl group on C6 has been positively identified as a pair of rapidly interconverting diastereomers while the adduct with the hydroxyl group on C8 is presumed to be a mixture of diastereomers since C8 is a chiral center. The proposed reaction pathway for the 8-hydroxy adduct is a Michael addition of the N2-amino group from the deoxyguanosine to the C3 of acrolein and the closing of the ring by a reaction between the N1 of deoxyguanosine and C1 of acrolein. The reaction pathway for the 6-hydroxy adduct is also proposed to involve a Michael addition, starting with a reaction between N1 of deoxyguanosine and the acrolein C3, followed by ring closure between the acrolein C1 and the deoxyguanosine N2.

The adduct(s) formed after the reaction of acrolein with deoxycytidine is not so clear (Fig. 5b). Using a combination of HPLC and ^{32}P postlabeling, the major adduct formed from the reaction of acrolein and deoxycytidine 5′-monophosphate was a mixture of

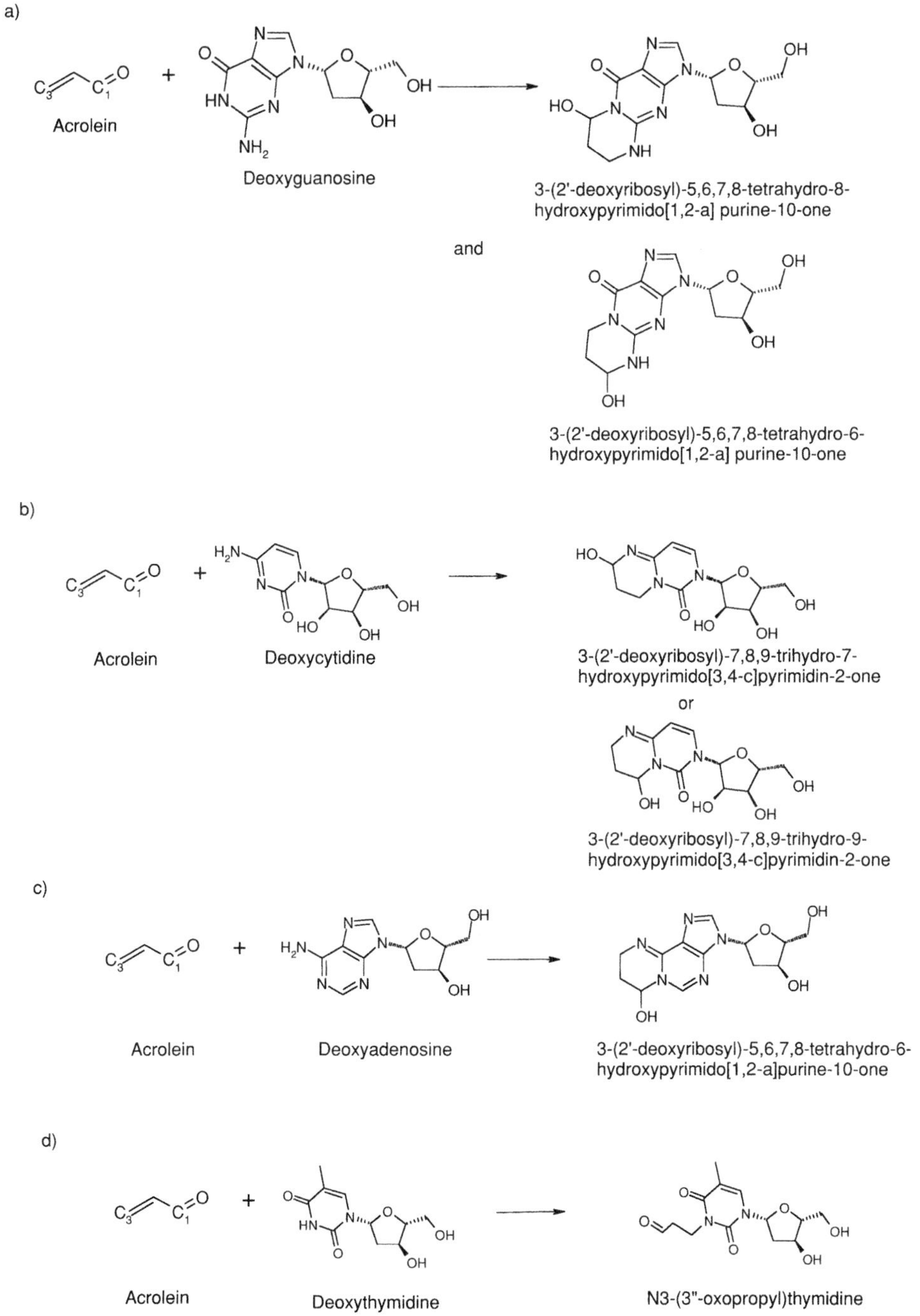

Fig. 5. Acrolein DNA adducts resulting from the reaction of acrolein and (a) deoxyguanosine, (b) deoxycytidine, (c) deoxyadenosine, and (d) deoxythymidine.

diastereomers identified as 3-(5′-monophospho-2′-deoxyribosyl)-7,8,9-trihydro-7-hydrox-yprimido[3,4-c]pyrimidin-2-one [87]. The proposed reaction pathway is the addition of C1 of acrolein to the N4-amino group position of cytidine in a Michael addition, followed by ring closure between the C3 of acrolein and N3 of the pyrimidine ring. Another group, using HPLC and with UV, ^{1}H NMR, and mass spectrometry techniques identified the same pair of diastereomers as the major adduct [88]. However, they proposed a different mechanism: addition of C3 of acrolein to N3 of deoxycytidine in a Michael addition followed by cyclization by the reaction of the exocyclic amine N4 of deoxycytidine to C1 of acrolein. Inconsistent with the above reports, another cytidine adduct, 3-(2′-deoxyri-bosyl)-7,8,9-trihydro-9-hydroxyprimido[3,4-c]pyrimidin-2-one, has been identified using spectroscopy and chromatography techniques [89]. The proposed mechanism is Michael addition of the N4-amino group of deoxycytosine to C3 of acrolein followed by ring closure between C1 of acrolein and N3 of deoxycytidine. Chenna *et al.* report that they saw no evidence of this form of the adduct [88].

Sodum *et al.* also report on a single adduct resulting from the reaction of acrolein to deoxyadenine (Fig. 5c) [89]. This adduct, 3-(2′-deoxyribosyl)-5,6,7,8-tetrahydro-6-hydroxypyrimido[1,2-a]purine-10-one, is analogous to their reported deoxycytidine adduct (Fig. 5b). The proposed reaction pathway is a Michael addition of the N6-amino group of deoxyadenosine to acrolein C3 followed by cyclization. The authors note that two sites on the base could form a cyclic adduct. The one observed involving the Michael addi-tion and a second involving the carbonyl group of acrolein. Both forms were reported with the deoxyguanosine adducts by Chung *et al.* (see above) [73].

Finally, there has been one major adduct resulting from the reaction of acrolein with thymidine under physiological conditions [75]. These products were separated by HPLC and characterized by spectrometry and NMR techniques. One major product, N3′-(3″-oxopropyl)thymidine was found (Fig. 5d). However, this product was not stable under aqueous conditions.

9. Biological relevance of acrolein adducts

While acrolein can form stable adducts *in vitro* with deoxycytidine, deoxyguanosine, and deoxyadenine, the major adduct formed from the interaction of acrolein and DNA *in vitro* under physiological conditions is the adduct with the hydroxyl group at C-8 (1,N^2-propan-odeoxyguanosine) [73], suggesting that this adduct is the biologically relevant adduct. Subsequent studies support this idea. ^{32}P-postlabeling followed by a nucleotide chromato-graphy technique found this to be the major adduct found in human fibroblasts treated with acrolein [90]. This major adduct comigrated with the 1,N2-deoxyguanosine standard and reacted with an antibody against this exocyclic compound. The same group detected 1,N2-propanodeoxyguanosine in the DNA of peripheral blood lymphocytes of a dog treated with a therapeutic dose of cyclophosphamide. The same was found *in vitro* after incubation of calf thymus DNA incubated with acrolein [86]. These data taken together support the con-clusions that not only is 1,N2-deoxyguanosine the major adduct produced in DNA reacted with acrolein but it appears to be the major DNA adduct formed after acrolein and cyclophosphamide treatment in mammalian cell culture and lymphocytic DNA.

As discussed above, acrolein and other α,β-unsaturated aldehydes undergo 1,4-conjugated addition on nucleobases, primarily guanine [73,86]. Since the 1 and N2 positions of guanine are involved in base pairing, these exocyclic adducts obstruct Watson-Crick base pairing and thus are anticipated to be premutagenic lesions that block replication. Mutagenicity assays introducing the major 1,N2-propanodeoxyguanosine adduct into *Salmonella typhimurium* have provided some clues as to the mechanisms underlying acrolein-induced mutagenicity [65–67]. Acrolein has been shown to be mutagenic in two *Salmonella* tester strains, TA100 and TA104, which have GC and AT basepairs at the site of reversion respectively [91,92].

To determine if the cyclic deoxyguanosine adducts are at least partially responsible for the acrolein-induced mutagenicity seen in *S. typhimurium*, concentrations of 2′-propanodeoxyguanosine were measured using an immunoassay and compared to the induction of mutagenicity. This study confirmed that acrolein is mutagenic in TA100 and TA104 and showed that 2′-propanodeoxyguanosine adducts were formed in a dose dependent manner [74]. The relationship between the 2′-propanodeoxyguanosine adduct and mutagenicity has been confirmed in *S. typhimurium* [66] and mammalian Chinese hamster ovary cells [94].

10. Conclusion

As has been shown in this review, the nitrogen and phosphoramide mustard anticancer agents have a rich and complex DNA-alkylation chemistry that can be modulated by chemical modifications and conjugation to DNA-targeting groups. On-going research is elucidating the chemical factors underlying this complex alkylation chemistry. Another important aspect of the biological activity of the phosphoramide mustard prodrugs is the production of acrolein as a metabolic byproduct. Like nitrogen mustards themselves, acrolein has a complex DNA-alkylation chemistry with many possible alkylation sites on different nucleotide bases. Ultimately, detailed chemical knowledge of the nitrogen and phosphoramide mustards and their metabolic byproducts will allow the development of new alkylating agents with greatly increased efficiency for forming therapeutically active DNA interstrand crosslinks and fewer non-therapeutic, but still mutagenic, monofunctional adducts.

Acknowledgements
This work was carried out at Lawrence Livermore National Laboratory under contract W-7405-ENG-48 from the U.S. Department of Energy.

References

[1] M. Colvin, B.A. Chabner, Alkylating agents. In: B.A. Chabner, J.M. Collins, eds, *Cancer Chemotherapy: Principles and Practice* (pp. 276–313). J.B. Lippincott Company, Philadelphia, 1990.
[2] S.M. Ludeman, *Curr. Pharm. Des.* 5 (1999) 627–643.
[3] S.R. Rajski, R.M. Williams, *Chem. Rev.* 98 (1998) 2723–2793.
[4] M.E. Colvin, J.C. Sasaki, N.L. Tran, *Curr. Pharm. Des.* 5 (1999) 645–663.

[5] P. Brookes, *Mutat. Res.* 233 (1990) 3–14.

[6] P. Calabresi, B.A. Chabner. Antineoplastic agents. In: A.G. Gilman, T.W. Rall, A.S. Nies, P. Taylor, eds, *Goodman and Gilman's the Pharmacological Basis of Therapeutics*, eighth edn (pp. 1209–1263). Pergamon Press, Elmsford, 1990.

[7] J. Hansson, R. Lewensohn, U. Ringborg, B. Nilsson, *Cancer Res.* 47 (1987) 2631–2637.

[8] M. Colvin. The alkylating agents. In: B. Chabner, ed., *Pharmacologic Principles of Cancer Treatment* (pp. 276–308). Saunders, Philadelphia, 1982.

[9] L.F. Povirk, D.E. Shuker, *Mutat. Res.* 318 (1994) 205–226.

[10] P. Brookes, P.D. Lawley, *Biochem. J.* 80 (1961) 496–503.

[11] M.R. Osborne, P.D. Lawley, *Chem.-Biol. Interact.* 89 (1993) 49–60.

[12] A.E. Maccubbin, L. Caballes, G.B. Chheda, R.F. Struck, H.L. Gurtoo, *Biochem. Biophys. Res. Commun.* 163 (1989) 843–850.

[13] W.A. Denny. New developments in the use of nitrogen mustard alkylating agents as anticancer drugs. In: G.B. Jones, ed., *Advances in DNA Sequence-Specific Agents*, vol. 3 (pp. 158–178). JAI, Greenwich, 1998.

[14] V. Murray. A survey of the sequence-specific interaction of damaging agents with DNA: Emphasis on anti-tumor agents. In: K. Moldave, ed., *Progress in Nucleic Acid Research and Molecular Biology*, vol. 63 (pp. 367–415). Academic Press, New York, 2001.

[15] O. Kikuchi, A.J. Hopfinger, G. Klopman, *Biopolymers* 19 (1980) 325–340.

[16] K.W. Kohn, J.A. Hartley, W.B. Mattes, *Nucleic Acids Res.* 15 (1987) 10531–10549.

[17] W.B. Mattes, J.A. Hartley, K.W. Kohn, *Nucleic Acids Res.* 14 (1986) 2971–2987.

[18] P.B. Hopkins, J.T. Millard, J. Woo, M.F. Weidner, J.J. Kirchner, S.T. Sigurdson, S. Raucher, *Tetrahedron* 47 (1991) 2475–2489.

[19] J.A. Hartley, J.P. Bingham, R.L. Souhami, *Nucleic Acids Res.* 20 (1992) 3175–3178.

[20] A. Pullman, B. Pullman, *Q. Rev. Biophys.* 14 (1981) 289–380.

[21] J.A. Hartley, S.M. Forrow, R.L. Souhami, *Biochemistry* 29 (1990) 2985–2991.

[22] R.A. Pearlstein, S.K. Tripathy, R. Potenzone Jr, D. Malhotra, A.J. Hopfinger, G. Klopman, N. Max, *Biopolymers* 19 (1980) 311–324.

[23] P.M. Cullis, L. Merson-Davies, M.J. Sutcliffe, R. Weaver, *Chem Commun.* (1998) 1699–1700.

[24] J.B. Smaill, J.Y. Fan, P.V. Papa, C.J. O'Connor, W.A. Denny, *Anti-Cancer Drug Des.* 13 (1998) 221–242.

[25] P.R. Turner, L.R. Ferguson, W.A. Denny, *Anti-Cancer Drug Des.* 14 (1999) 61–70.

[26] J.Y. Fan, K.K. Valu, P.D. Woodgate, B.C. Baguley, W.A. Denny, *Anti-Cancer Drug Des.* 12 (1997) 181–203.

[27] A.S. Prakash, W.A. Denny, T.A. Gourdie, K.K. Valu, P.D. Woodgate, L.P.G. Wakelin, *Biochemistry* 29 (1990) 9799–9807.

[28] I.V. Kutyavin, H.B. Gamper, A.A. Gall, R.B. Meyer, *J. Am. Chem. Soc.* 115 (1993) 9303–9304.

[29] N. Brooks, P.J. McHugh, M. Lee, J.A. Hartley, *Chem. Biol.* 7 (2000) 659–668.

[30] M.D. Wyatt, M. Lee, B.J. Garbiras, R.L. Souhami, J.A. Hartley, *Biochemistry* 34 (1995) 13034–13041.

[31] M.D. Wyatt, B.J. Garbiras, M.K. Haskell, M. Lee, R.L. Souhami, J.A. Hartley, *Anti-Cancer Drug Des.* 9 (1994) 511–525.

[32] W.A. Denny, *Curr. Medicinal Chem.* 8 (2001) 533–544.

[33] A.S. Prakash, W.A. Denny, L.P.G. Wakelin, *Chem.-Biol. Interact.* 76 (1990) 241–248.

[34] J.N. Lampe, I.V. Kutyavin, R. Rhinehart, M.W. Reed, R.B. Meyer, H.B. Gamper, *Nucleic Acids Res.* 25 (1997) 4123–4131.

[35] H. Arnold, F. Bourseaux, N. Brock, *Arzneimittel-Forschung* 11 (1961) 143–158.

[36] T. Aida, W.J. Bodell, *Cancer Res.* 47 (1987) 1361–1366.

[37] M. Gniazdowski, C. Cera, *Chem. Rev.* 96 (1996) 619–634.

[38] J.P.H. Wijen, M.J.M. Nivard, E.W. Vogel, *Carcinogenesis* 21 (2000) 1859–1867.

[39] F.H. Hausheer, U.C. Singh, J.D. Saxe, O.M. Colvin, *Anti-Cancer Drug Des.* 4 (1989) 281–294.

[40] J.T. Millard, S. Raucher, P.B. Hopkins, *J. Am. Chem. Soc.* 112 (1990) 2459–2460.

[41] J.O. Ojwang, D.A. Grueneberg, E.L. Loechler, *Cancer Res.* 49 (1989) 6529–6537.

[42] P.M. Cullis, R.E. Green, M.E. Malone, *J. Chem. Soc. Perkin Trans.* 2 (1995), 1503–1511.

[43] Q. Dong, D. Barsky, M.E. Colvin, C.F. Melius, S.M. Ludeman, J.F. Moravek, O.M. Colvin, D.D. Bigner, P. Modrich, H.S. Friedman, *Proc. Natl. Acad. Sci. USA* 92 (1995) 12170–12174.

[44] S.M. Rink, M.S. Solomon, M.J. Taylor, S.B. Rajur, L.W. McLaughlin, P.B. Hopkins, *J. Am. Chem. Soc.* 115 (1993) 2551–2557.

[45] S.M. Rink, P.B. Hopkins, *Biochemistry* 34 (1995) 1439–1445.
[46] R.T. Streeper, R.J. Cotter, M.E. Colvin, J. Hilton, O.M. Colvin, *Cancer Res.* 55 (1995) 1491–1498.
[47] G.B. Bauer, L.F. Povirk, *Nucleic Acids Res.* 25 (1997) 1211–1218.
[48] M.R. Osborne, D.E.V. Wilman, P.D. Lawley, *Chem. Res. Toxicol.* 8 (1995) 316–320.
[49] R.M. Romero, P. Rojsitthisak, I.S. Haworth, *Arch. Biochem. Biophys.* 386 (2001) 143–153.
[50] M. Brendel, A. Ruhland, *Mutat. Res.* 133 (1984) 51–85.
[51] A. Mudipalli, A.E. Maccubbin, S.S. Nadadur, R.F. Struck, H.L. Gurtoo, *Mutat. Res.-Fundam. Mol. Mech. Mut.* 381 (1997) 49–57.
[52] B.M. Yaghi, P.M. Turner, W.A. Denny, P.R. Turner, C.J. O'Connor, L.R. Ferguson, *Mutat. Res.-Fundam. Mol. Mech. Mut.* 401 (1998) 153–164.
[53] Y.N. Cai, M.H. Wu, M. Xu-Welliver, A.E. Pegg, S.M. Ludeman, M.E. Dolan, *Cancer Res.* 60 (2000) 5464–5469.
[54] M.E. Dolan, S.K. Roy, A.A. Fasanmade, P.R. Paras, R.L. Schilsky, M.J. Ratain, *J. Clin. Oncol.* 16 (1998) 1803–1810.
[55] R.F. Struck, R.L. Davis, M.D. Berardini, E.L. Loechler, *Cancer Chemother. Pharmacol.* 45 (2000) 59–62.
[56] Y.H. Fan, B. Gold, *J. Am. Chem. Soc.* 121 (1999) 11942–11946.
[57] J.T. Millard, M.M. White, *Biochemistry* 32 (1993) 2120–2124.
[58] M.J. Yunes, S.E. Charnecki, J.J. Marden, J.T. Millard, *Chem. Res. Toxicol.* 9 (1996) 994–1000.
[59] J.K. Strauss, C. Roberts, M.G. Nelson, C. Switzer, L.J. Maher, *Proceedings of the National Academy of Sciences of the United States of America* 93 (1996) 9515–9520.
[60] M.G. Remias, C.-S. Lee, I.S. Haworth, *J. Biomol. Struct. Dyn.* 12 (1995) 911–936.
[61] C.J.M. Stirling, *Journal of Chemical Education* 50 (1973) 844–845.
[62] R. Bird, A.C. Knipe, C.J.M. Stirling, *Journal of the Chemical Society Perkins Transactions II* (1973) 1215–1220.
[63] S. Ludeman, *Curr. Pharm. Des.* 5 (1999) 627–643.
[64] G. Eisenbrand, J. Schuhmacher, P. Gölzer, *Chem. Res. Toxicol.* 8 (1995) 40–46.
[65] L.S. Marnett, H.K. Hurd, M.C. Hollstein, D.E. Levin, H. Esterbauer, B.N. Ames, *Mutat. Res.* 148 (1985) 25–34.
[66] E. Eder, C. Hoffman, H. Bastian, C. Deininger, S. Scheckenbach, *Environmental Health Perspectives* 88 (1990) 99–106.
[67] D. Lutz, E. Eder, T. Neudecker, D. Henschler, *Mutat. Res.* 93 (1982) 305–315.
[68] L.M. Sierra, A.R. Barros, M. Garcia, J.A. Ferreiro, M.A. Comendador, *Mutat. Res.* 260 (1991) 247–256.
[69] C. Monteil, E. Le Prieur, S. Buisson, J.P. Morin, M. Guerbet, J.M. Jouany, *Toxicology* 133 (1999) 129–138.
[70] J.P. Kehrer, S.S. Biswal, *Toxicological Sciences* 57 (2000) 6–15.
[71] E. Agostinelli, E. Przybytkowski, D.A. Averill-Bates, *Free Radical Biol. Med.* 20 (1996) 649–656.
[72] J.M. Patel, E.R. Block, *Experimental Lung Research* 8 (1985) 153–165.
[73] F.L. Chung, R. Young, S.S. Hecht, *Cancer Res.* 44 (1984) 990–995.
[74] P.G. Foiles, S.A. Akerkar, F.L. Chung, *Carcinogenesis* 10 (1989) 87–90.
[75] A. Chenna, R.A. Rieger, C.R. Iden, *Carcinogenesis* 13 (1992) 2361–2365.
[76] R. Shapiro, R.S. Sodum, D.W. Everett, S. Kundu, Reactions of nucleosides with glyoxal and acrolein. In: B. Singer, H. Bartsch, eds, *The Role of Cyclic Nucleic Acid Adducts in Carcinogenesis and Mutagenesis* (pp. 165–173). IARC Scientific Publications 70, Lyon, 1986.
[77] T.R. Crook, R.V. Souhami, A.E.M. McLean, *Cancer Res.* 46 (1986) 5029–5034.
[78] R.C. Grafström, J.M. Dypbukt, J.C. Willey, K. Sundqvist, C. Edman, L. Atzori, C.C. Harris, *Cancer Res.* 48 (1988) 1717–1721.
[79] T.M. Harris, I.D. Kozekov, L.V. Nechev, A. Sanchez, C.M. Harris, R.S. Lloyd. In: *Meeting of the American Chemical Society*, vol. 222 (p. TOXO 20). American Chemical Society, Washington DC, 2001.
[80] T.M. Harris, I.D. Kozekov, L.V. Nechev, A. Sanchez, C.M. Harris, R.S. Lloyd. In: *Meeting of the American Chemical Society*, vol. 222 (p. TOXI 24). American Chemical Society, Washington DC, 2001.
[81] L.A. VanderVeen, M.F. Hashim, L.V. Nechev, T.M. Harris, C.M. Harris, L.J. Marnett, *Journal of Biological Chemistry* 276 (2001) 9066–9070.
[82] B. Singer, H. Bartsch, eds, *Exocyclic DNA Adducts in Carcinogenesis and Mutagenesis*, IARC Scientific Publications 150, Lyon, 1999.

[83] M. Moriya, E. Marinelli, S. Shibutani, J. Joseph. In: *American Association for Cancer Research Annual Meeting*, vol. 30 (p. 140). American Association for Cancer Research, 1989.

[84] S.S. Hecht, Carcinogen biomarkers for lung or oral cancer chemoprevention trials. In: A.B. Miller, H. Bartsch, P. Boffetta, L. Dragsted, H. Vaino, eds, *Biomarkers in Cancer Chemoprevention* (pp. 245–255). IARC Scientific Publications 154, Lyon, 2001.

[85] M.P. Gamcsik, M.E. Dolan, B.S. Andersson, D. Murray, *Current Pharmaceutical Design* 5 (1999) 587–605.

[86] T. Douki, B.N. Ames, *Chem. Res. Toxicol.* 7 (1994) 511–518.

[87] R.A. Smith, D.S. Williamson, S.M. Cohen, *Chem. Res. Toxicol.* 2 (1989) 267–271.

[88] A. Chenna, C.R Iden, *Chem. Res. Toxicol.* 6 (1993) 261–268.

[89] R.S. Sodum, R. Shapiro, *Bioorganic Chemistry* 16 (1988) 272–282.

[90] V.L. Wilson, P.G. Foiles, F.L. Chung, A.C. Povey, A.A. Frank, C.C. Harris, *Carcinogenesis* 12 (1991) 1483–1490.

[91] D.M. Maron, B.N. Ames, *Mutat. Res.* 113 (1983) 173–215.

[92] D.E. Levin, M. Hollstein, M.F. Christman, E.A. Schwiers, B.N. Ames, *Proceedings of the National Academy of Sciences, USA* 79 (1982) 7445–7449.

[93] R.A. Smith, S.M. Cohen, T.A. Lawson, *Carcinogenesis* 11 (1990) 497–498.

[94] P.G. Foiles, S.A. Akerkar, L.M. Miglietta, F.L. Chung, *Carcinogenesis* 11 (1990) 2059–2061.

Advances in DNA Sequence-specific Agents 04 (2002) 47–73

Molecular basis for recognition and binding of specific DNA sequences by calicheamicin and duocarmycin

Giuseppe Bifulco[a], Jarrod A. Smith[b], Walter J. Chazin[b]*
and Luigi Gomez-Paloma[a]**

[a] *Dipartimento di Scienze Farmaceutiche, Università di Salerno, via Ponte don Melillo, 84084 Fisciano (SA), Italy. (39) (089) 962811, FAX: 962828.*

[b] *Department of Biochemistry & Center for Structural Biology, Vanderbilt University, Preston Building — 896, Nashville, TN 37232–0146, USA. (615) 936–2209, FAX: 936–2211*

1. Introduction

Molecules known to bind the minor groove of DNA have frequently been shown to exert cytotoxic activity by interfering with the binding of proteins necessary for DNA replication, transcription, and repair. Current chemotherapy is largely based on drugs with poor selective cytotoxicities, and this has motivated researchers to design new molecules that can bind selectively in the minor groove of DNA. Minor groove binding ligands with increased selectivity should produce a greater pharmacological response for a given dose than non-selective binders.

2. Calicheamicin

A. The enediyne family

The enediyne family of antitumor antibiotics are characterized by a unique molecular architecture, an intriguing mode of action, and highly potent biological and pharmacological activity [1]. These compounds exert strong and selective DNA damage by

*Corresponding author.
E-mail address: walter.chazin@vanderbilt.edu
**Corresponding author.
E-mail address: gomez@unisa.it
Advances in DNA Sequence-specific Agents, Volume 4 G. B. Jones (Editor)

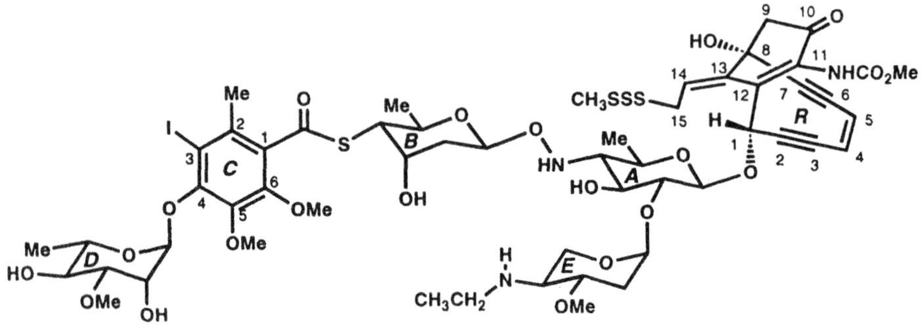

1

Fig. 1. Chemical structure of calicheamicin γ_1^1.

generating a diradical species capable of cleaving the DNA-duplex. In the last decade, a number of natural and synthetic molecules belonging to this family have been investigated in an effort to fully understand the mechanism of cleavage, and to provide insight into the factors governing their affinity and binding selectivity for duplex DNA.

Enediynes possess a common reactive center along with a chemical functionality that can trigger a cascade of events to form a highly reactive aryl-diradical. These molecules also possess a "clasp" device, responsible for delivering and securing their position to the biological target, i.e. DNA.

When the molecule is bound to the DNA duplex, the trigger functionality is activated by external nucleophilic attack to give an intramolecular Michael addition followed by a Bergman cyclization which transforms the enediyne moiety into a very reactive aryl-diradical. The diradical goes on to abstract hydrogen atoms from each DNA strand leading to scission of the duplex. This damage to the DNA is not readily repaired by any of the known DNA repair pathways.

The calicheamicin family possesses activity against Gram-positive and Gram-negative bacteria and also shows activity (at doses in the range of 100ng/kg) against murine leukemias, solid neoplasms and melanomas [1]. Calicheamicin (**1**) has recently proven to be a potential lead compound for chemotherapy based on the finding that it induces apoptosis [2]. Inducers of apoptosis can have an important role in cancer therapy, in AIDS therapy and in the treatment of organ transplants. Calicheamicin (**1**) possesses an oligosaccharide delivery system which binds in the minor groove of the DNA. When the molecule has been delivered to duplex DNA, its triggering device, a trisulfide group, is activated by an attack of a strong nucleophile from the cellular environment (e.g. glutathione), forming a thiolate that is in an optimal position to attack the α,β-unsaturated ketone incorporated in the adjacent six-membered ring [3]. The hybridization of the carbon attacked by the thiolate is converted from sp^2 to sp^3, and causes the formation of a highly strained intermediate. This intermediate, characterized by a shortened distance between the ending atoms of the enediyne system (3.16 Å *vs.* the 3.35 Å of the intact calicheamicin), is ready for a Bergman cyclization reaction that will eventually transform the intermediate into a benzenoid diradical.

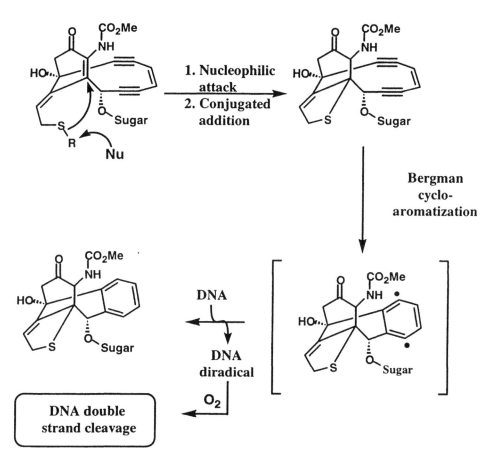

Fig. 2. Mechanism for DNA cleavage by calicheamicin γ_1^{I}.

B. Structure–activity relationships

Among the enediynes, calicheamicin γ_1^{I} (**1**) is one of the most interesting; it exhibits preferential binding to homopyrimidine tracts, particularly TTTT, ACCT, TCCT and TCTC. Footprinting studies on a series of synthetic DNA duplexes containing the binding site TCCT indicated that calicheamicin γ_1^{I} (**1**) anchors itself on the DNA from the 5′ side of the TCCT sequence and cuts double stranded helical DNA in close proximity to the target site in a manner reminiscent of a restriction endonuclease [4].

It became possible to assess the importance of various structural features in the carbohydrate portion of calicheamicin by examining synthetic derivatives. Removal of the 4-ethylamino sugar or acetylation of the 4-ethylamino group did not affect the specificity, which was similar to that of the parent compound, but caused a drop in the efficiency of the cleavage of DNA by two or three orders of magnitude. This suggested a catalytic role

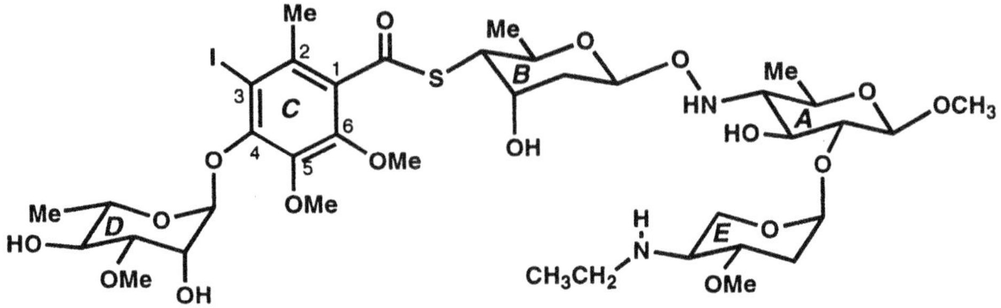

2

Fig. 3. Chemical structure of the calicheamicin oligosaccharide.

for the basic 4-ethylamino sugar (ring E of Fig. 1) in the activation of the trisulfide group. Studies on a derivative lacking of the terminal rhamnose (ring D of Fig. 1) showed the same sequence selectivity but a lower DNA scission efficiency. The combined removal of the 4-ethylamino sugar and the terminal rhamnose afforded a derivative that lacked the ability to cleave DNA.

Calicheamicin (**1**) binds to the target sequences with relatively high affinity. In a quantitative footprinting study using DNA plasmids, the K_D for calicheamicin and duplexes containing the preferred consensus site d(TCCT)·(AGGA) was determined to be 10 nM [5]. More recent measurements of the K_D of calicheamicin for a range of high affinity sequences showed this value is in the micromolar range (0.1–1.0 μM, see below).

Footprinting experiments, conducted with synthetic oligosaccharide domains lacking the aglycone, showed that the calicheamicin oligosaccharide (**2**) also binds specifically to DNA. Interestingly the preferred binding sites do not coincide precisely with those observed for intact calicheamicin. These results suggest that a significant portion of the binding specificity is due to the carbohydrate domain, but that the enediyne moiety confers added specificity and increases the affinity to duplex-DNA [6].

In order to gain further insight into the molecular basis for recognition and cleavage of DNA by calicheamicin, Gomez-Paloma et al. examined the complex between the d(GCATCCTAGC)·d(GCTAGGATGC) and calicheamicin γ_1^I (**1**) as well as the corresponding complex with the methyl glycoside of the oligosaccharide moiety (**2**) [7]. It is noteworthy that calicheamicin does not react with DNA unless exogenous thiol is present. Molecular mechanics and molecular dynamics calculations were carried out with NMR-derived distance restraints and afforded two medium resolution structures for these ligand-DNA complexes. The analysis of the NMR data and of the structures relative to the two complexes showed very close global and local conformational features, suggesting that the oligosaccharide domain of the calicheamicin is the main structural element in the control of binding and sequence recognition of the DNA minor groove. Nevertheless, the presence of the aglycone increases the affinity of the sugar toward DNA, probably due to the lipophilic character of the aglycone [8].

Determination of oligosaccharide-DNA binding constants was possible by comparing the known K_D of calicheamicin and the K_I (inhibition) for the binding of the oligosaccharide in competition with the parent compound (μM range vs. 100 nM). The proposed interaction of the iodine atom on ring C with the exocyclic amino groups of the guanines in the d(AGGA)·(TCCT) tracts [9] was examined by preparing a series of synthetic calicheamicin aryltetrasaccharides differing in functional groups at this position. These binding studies revealed that an iodine atom is crucial for binding to this site, as binding affinities decrease in the order I > Br > Cl > F > CH_3 > H [8]. The guanine 2-amino group within the DNA binding site has also been proven to be a critical recognition element, since it has been shown that replacing the guanine residue with inosine (which lacks guanine's 2-amino group but is otherwise identical) causes a dramatic reduction of binding affinity and selectivity towards calicheamicin [10].

C. Structures of calicheamicin-DNA complex: study of ligand-DNA interactions at the atomic level

Kahne and co-workers undertook the first NMR studies on the interaction of calicheamicin (**1**) with DNA [11]. They were able to obtain a stable complex between calicheamicin and the DNA duplex oligo d[GTGACCTTG]–d[CAGGTCAC], containing the ACCT recognition sequence. Their analysis of the NMR data indicated that the global conformation of calicheamicin does not change significantly upon binding. However, some of the experimental data showed that the C6 deoxyribose changes significantly upon binding of calicheamicin. In accord with the proposal of Zein and co-workers [4]. Walker *et al.* proposed a model in which calicheamicin recognizes DNA duplexes that have a wider than average minor groove, allowing an induced fit mode of binding [12]. In this model, binding site selectivity of calicheamicin is due to the ability of pyrimidine/purine tracts to accommodate the particular shape of the ligand.

Careful analysis of the three dimensional structure obtained by Gomez-Paloma *et al.* showed a set of interesting Van der Waals contacts and numerous hydrogen bond interactions for both calicheamicin (**1**) and its aryl tetrasaccharide (**2**) (Fig. 4). The aromatic ring C is found to stack with the deoxyribose ring of the A17. This type of aromatic interaction has been observed in other ligand-DNA complexes, and may also involve participation of the ring oxygen of G16. A strong salt bridge is formed between the positively charged nitrogen of the ethylamino group of calicheamicin ring E and the phosphate oxygen of C5. A hydrogen bond is observed between the 2-hydroxyl proton of the D-ring and a backbone oxygen atom of A17. These interactions are not dependent on the binding sequence and are considered important for the overall stabilization of the complex. The structures of these complexes reveal additional interactions that are strictly related to the sequence of the binding site, which are therefore expected to contribute to the sequence selectivity of calicheamicin. Among these interactions, two hydrogen bonds are particularly noteworthy: one involves the 3-hydroxyl proton of ring B and N3 of the A17 and the other the carbonyl oxygen of the ligand and the N2 amino group of G16. The importance of the iodine atom, described by Schreiber [9] and discussed above (in the context of the lower binding affinity of derivatives containing substituents other than

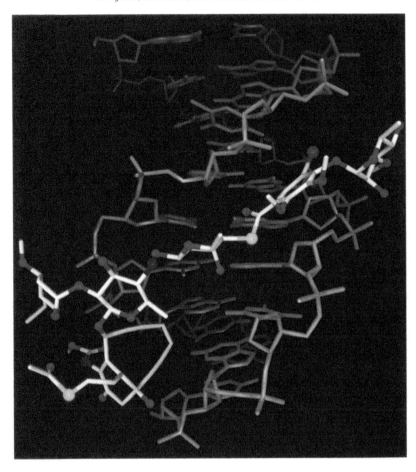

Fig. 4. Three-dimensional structures of calicheamicin γ_1^I in complex with d-(GCA*TCCT*AGC)·d-(GCT*AG-GA*TGC). (a) Overview of the complex with the two strands of DNA in pink and blue, and the ligand represented in white with colored spheres for the heteroatoms (red = O, blue = N, yellow = S, purple = I). (b) Critical inter-molecular interactions in the binding site of the high resolution structure of the isolated oligosaccharide domain. Three hydrogen bonds are depicted as white dotted lines, and a salt bridge as a green dotted line. A space-filling representation of a soft hydrogen bonding interaction between the iodine atom and the amino group of G15 is also shown.

iodine on the C ring) is confirmed by the evidence of a contact between the iodine atom, which acts as a soft hydrogen bond acceptor, and the G15 N2 amino group.

D. Role of the hydroxylamine glycosidic linkage

The large majority of the minor groove binding DNA ligands adopt a shape that is complementary to the curvature of the DNA minor groove. The presence of peptidyl

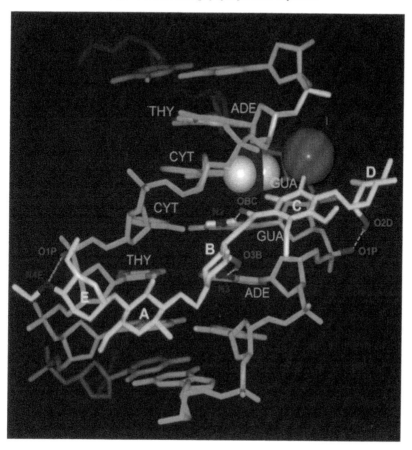

Fig. 4. *(cont.)*

linkages, as observed for instance in netropsin, distamycin, and duocarmycin, is often responsible for the flexibility that is needed to obtain correct curvature for binding. The peculiar hydroxylamine glycosidic linkage that connects rings A and B is one key element that allows calicheamicin to have the proper shape necessary to accommodate the DNA minor groove and enable an optimal fit of the two parts of the molecule [13]. Although it has a very different chemical nature, this linkage can be considered a structural homolog to the peptidyl linkage.

E. Oligomeric saccharides as new minor groove binders

The possibility of synthesizing minor groove binding agents that are selective for AT and GC base pairs by linking two or more DNA binding agents was first explored by Dervan and Lown [14,15]. They have prepared a number of conjugated molecules based on the dystamycin/netropsin scaffolds with significant sequence selectivity. Netropsin,

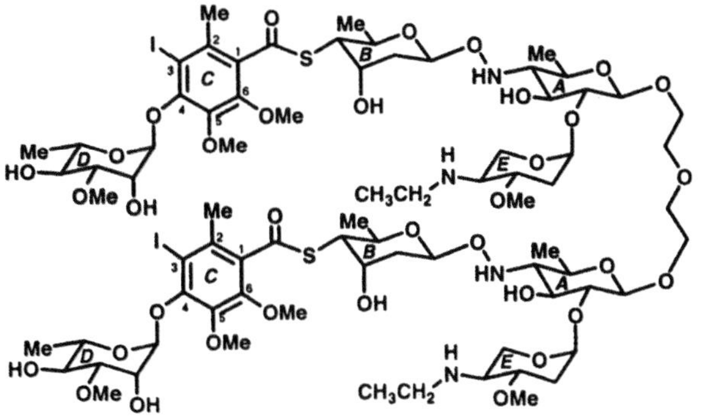

3

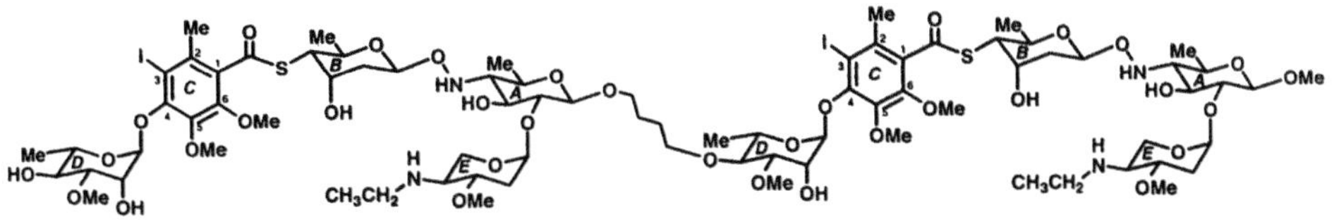

4

Fig. 5. Chemical structures of head-to-head (3) and head-to-tail (4) calicheamicin oligosaccharide dimers.

known to bind preferentially in A-T sequences, was modified to allow the recognition of a G-C base pair [15] by replacing a pyrrole ring in netropsin by an *N*-methylimidazole unit. This modification enabled recognition of the amino groups in the minor groove of C-G pairs. The combination of pyrrole-imidazole polyamides, characterized by a simple molecular shape, can achieve affinities and specificities comparable to DNA-binding proteins and have the potential to target any desired DNA sequence.

The same concepts have been used to prepare new oligomeric DNA binding agents based on the oligosaccharide moiety of calicheamicin. Two dimeric forms of calicheamicin oligosaccharide, head-to-head (3) and head-to-tail (4) dimers were designed [16] on the basis of information obtained from the three dimensional structure of the complex between the calicheamicin aryl tetrasaccharide and DNA.

The selectivity of these two dimers for the designed binding sites has been confirmed by competitive binding experiments conducted using calicheamicin as a cleaving agent; affinities in the low nanomolar range have been deduced for head-to-head and head-to-tail dimers, respectively 100-fold and 1000-fold higher than the monomeric oligosaccharide [17].

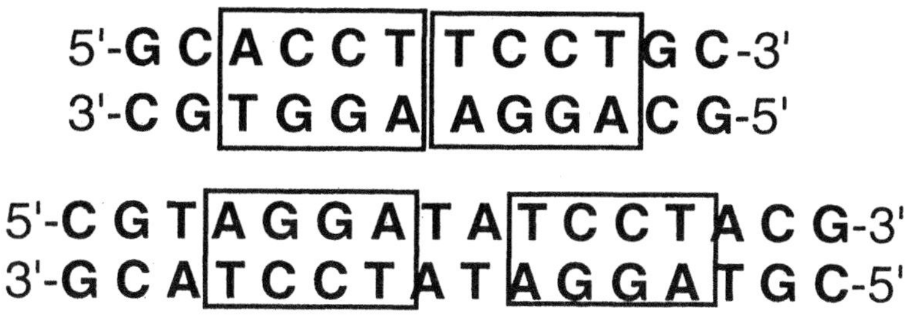

5'-G C $\boxed{\text{A C C T}}$ $\boxed{\text{T C C T}}$ G C-3'
3'-C G $\boxed{\text{T G G A}}$ $\boxed{\text{A G G A}}$ C G-5'

5'-C G T $\boxed{\text{A G G A}}$ T A $\boxed{\text{T C C T}}$ A C G-3'
3'-G C A $\boxed{\text{T C C T}}$ A T $\boxed{\text{A G G A}}$ T G C-5'

Fig. 6. DNA sequences used in the complexes of the head-to-head (upper panel) and head-to-tail (lower panel) calicheamicin oligosaccharide dimers. The recognition sites are enclosed in boxes.

The higher selectivity and affinity of the carbohydrate dimers has prompted a study of the inhibition of the binding of transcription factors to sequences containing the binding site (i.e. TCCT) by the head-to-head dimer (3). These studies have confirmed that carbohydrates can antagonize the sequence-specific action of transcription factors at a concentration in the micromolar range and that the dimer of the aryltetrasaccharide is a much more effective agent in transcriptional inhibition, being 10 times more active than the calicheamicin oligosaccharide [18].

The head-to-head (3) and the head-to-tail (4) dimers of calicheamicin oligosaccharide have an impressive nanomolar affinity for their specific DNA consensus sites, and can selectively recognize sequences containing eight base pairs. High resolution structures of these dimers complexed to DNA have been determined in order to understand the molecular basis for the high DNA affinity of these molecules. A structure has been obtained for the complex between the head-to-head dimer and a self-complementary sixteenmer, d(CGTAGGATATCCTACG)₂ [19]. This DNA duplex contains two TCCT recognition sites, separated by two base pairs that don't actively participate in the binding but which are inserted to span the region of the bis(ethylene-glycol) linker that connects the two monomeric saccharide subunits.

The head-to-head dimer (3) binds to the DNA duplex in two subtly different conformations attributed to differences in the positioning of ring E and E'. A careful analysis of the structure revealed favorable interactions that account for the binding sequence preferences and the stability of the complex. The interactions observed for each of the oligosaccharide subunits of this dimer seem to reproduce those observed for the calicheamicin oligosaccharide monomer bound to a single TCCT recognition site. The studies of the head-to-head dimer showed that the design objectives for the dimeric system were met and suggested that carbohydrates may be useful as a means for generating molecules with high DNA target specificity and affinity.

The study of the head-to-head dimer was complemented by corresponding analysis of the head-to-tail dimer (4), using the DNA duplex d(GCACCTTCCTGC)–d(GCAGGAA GGTGC) with one d(ACCT) and one d(TCCT) high affinity binding site [20].

Table 1
HHD–DNA intermolecular H-bonds[a]
observed in the solution structure for the
1:1 HHD–DNA complex

Donor	Acceptor
B–3OH	A_{23}–N3
G_{22}–NH_2	C–SC=O
D–2OH	A_{23}–O1 P
B'–3OP	A_7–N3
G_6–NH_2	C'–SC=O
D'–2OH	A_7–O1 P

[a]The H-bonds are defined with a distance cutoff of 3.0 Å.

In contrast to the head-to-head dimer (**3**), the head-to-tail dimer (**4**) exhibits a unique binding mode in the DNA minor groove, reflecting an improved molecular design. A comparative analysis of the carbohydrate-DNA interactions at the two different binding sites showed how the calicheamicin carbohydrate is able to effectively recognize both d(ACCT) and d(TCCT) sites. The structure revealed intermolecular interactions between the DNA and the dimer that were very similar to those observed for the isolated subunits. In comparing the two different sites, one finds a hydrogen bond formed from the carbo-hydrate B-3OH of each subunit: a thymine O2 was used in the ACCT binding site (5'-ACCT-3'/5'-AGGT_{22}-3') as opposed to the adenine N3 used in the TCCT site (5'-TCCT-3'/5'-AGGA_{16}-3'). These structural studies have highlighted the importance of both specific and non-specific interactions mediating the binding of the calicheamicin oligo-saccharide in the DNA minor groove. There is an ongoing debate regarding the relative importance of adaptability of the DNA sequence versus specific interactions (that will vary for the four high affinity sites) in determining the sequence preferences of the calicheamicin oligosaccharide. Comparative analysis of high resolution three-dimensional structures for all four high affinity sites is required to fully appreciate the molecular basis for the selective recognition of DNA by calicheamicin.

3. Duocarmycin

A. The family of CC-1065 and duocarmycins: an introduction

The duocarmycins and CC-1065 are the parent members of a class of exceptionally potent antitumor antibiotics that exert their biological activity through sequence selective alkylation of duplex DNA [21]. These compounds display *in vitro* IC_{50} against tumor cell lines at the pM level, and certain derivatives exhibit antitumor selectivity. For example,

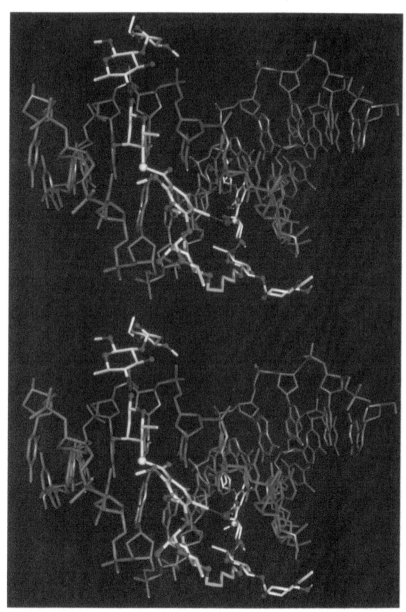

Fig. 7. Stereoview of the three-dimensional structure of calicheamicin oligosaccharide head-to-tail dimer bound to d(CGT<u>AGGA</u>TA<u>TCCT</u>ACG)$_2$. The two DNA strands are drawn in blue and orange. The two ligand subunits are drawn in silver and gold, with the linker between them in green.

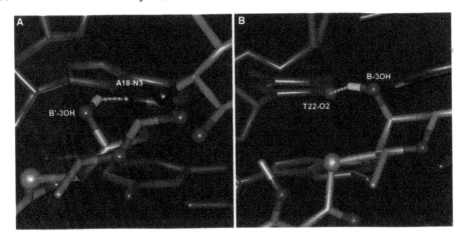

Fig. 8. Comparison of hydrogen bonding of ring B at the ACCT versus TCCT binding sites in the complex of the calicheamicin head-to-tail dimer bound to d(CGTAGGATATCCTACG)₂.

duocarmycin SA (DSA) is able to cause apoptosis (programmed cell death) in tumor cells at a dose (<100 pM) well below that required for cell death by necrosis in non-sensitive cell lines [22]. Compounds under investigation for clinical use include the CC-1065-derived cross-linking agent bizelesin, which is currently in phase I clinical trials [23] and KW-2189, a derivative of duocarmycin B2, which is undergoing clinical trials in Japan [24]. In addition to their potential for direct use as anticancer agents, duocarmycins have also been proposed as conjugates for antibody-directed chemotherapy [25]. Their potent bioactivity and intriguing mechanism of action have attracted a wealth of chemical and pharmacological investigations that have resulted in the preparation of a number of derivatives [26], in the clarification of the molecular basis of their DNA alkylating properties [27], and ultimately in the synthesis of tailored bioconjugates that appear to be promising agents for chemotherapy of tumors [28].

Investigations of the chemistry and biology of these agents commenced in 1978, when CC-1065 was isolated from cultures of *Streptomyces zelensis* at Upjohn and *Streptomyces C-329* at Brystol-Meyers [29]. The structure of this remarkable natural product was initially deduced from spectroscopic evidence in 1980 and then secured by X-ray analysis [30]. It consists of two identical pyrrolo[3,2-e] indole subunits (the right-hand and central subunits depicted in structure 5, usually referred to as *binding subunits*) connected through an amide linkage to a further pyrrolo[3,2-e]indole system that bears a spirocyclic cyclopropane ring (the left-hand portion of 5, usually referred to as *alkylating subunit*). Soon after its identification, CC-1065 was shown to be a very active cytotoxin with an IC₅₀ approximately three orders of magnitude lower than other agents known at that time [31].

Duocarmycin A (6) and the more chemically stable duocarmycin SA (7) were isolated in 1988 and 1990, respectively, from Japanese strains of *Streptomyces* [32]. These compounds, characterized by the presence of only a single binding subunit, proved to be particularly interesting, both for their increased stability and improved pharmacological properties [33], including the lack of the delayed fatal toxicity observed after *in vivo* treatments with CC-1065 [34].

Extensive efforts have been devoted to understanding of the binding preference of these ligands towards the minor groove of AT-rich tracts of DNA, the details of the alkylation reaction with duplex DNA, and the source of catalysis [35]. The alkylation reaction is a reversible, stereoelectronically controlled addition by N3 of adenine to the least substituted carbon of the cyclopropane ring. The reaction with small nucleophiles is rather slow, but a vast acceleration is observed when the agent is bound to the minor groove of AT-rich sequences of DNA (see section 2.3). The origin of the DNA alkylation selectivity is believed to lie in the intrinsic (noncovalent) binding preference of duocarmycins towards AT-rich DNA sequences (noncovalent binding model), although other models have also been proposed (see section 2.2).

The sequence selectivity of duocarmycin parallels (at least in a general sense) that observed for other pseudopeptidic ligands, such as netropsin and distamycin. All these gently curved, flat, multi-subunit heteroaromatic molecules are characterized by their ability to follow the helical pitch of duplex DNA through appropriate inter-ring linkages. They each exhibit a hydrophobic inner surface and a hydrophilic outer surface. Significant Van der Waals interactions are anticipated with the floor and the walls of the deep and narrow minor groove of AT-rich tracts.

A correlation exists between structural features governing sequence-selective DNA alkylation and biological activity. Initial expectation of a direct relationship between the reactivity of the cyclopropyl functionality and cytotoxic activity suggested that enhancement of the electrophilic reactivity of the ring should lead to increased biological potencies [36]. However, studies over the past 10 years by Boger and co-workers have shown that the most stable compounds exhibit the most potent cytotoxicities. A well defined, direct linear relationship between stability towards solvolysis and biological potency has been observed. This relationship holds for both simplified and full analogs of the parent compound. Similarly, a linear relationship was found between the electron-withdrawing properties of the N2 substituents (Hammet σ_p constant) and solvolysis reactivity ($-\log k$ at pH = 3) in which the strongest electron-withdrawing substituents provide the most stable derivatives [37]. This can be attributed to the influence of N2 substituents on the C4 protonation required for the acidic solvolysis. Combining together these two linear relationships, cytotoxicity and the electron-withdrawing power of N2 substituents were found to be linearly related [38].

The relationship between relative reactivity of the cyclopropyl function and cytotoxicity can be explained in general terms by assuming that better stability (lower intrinsic reactivity) ensures more effective delivery of the agents to their intracellular target. In other words, compounds possessing a high degree of electrophilic character can be trapped by other cellular nucleophiles reducing the amount of agent available for the desired alkylation.

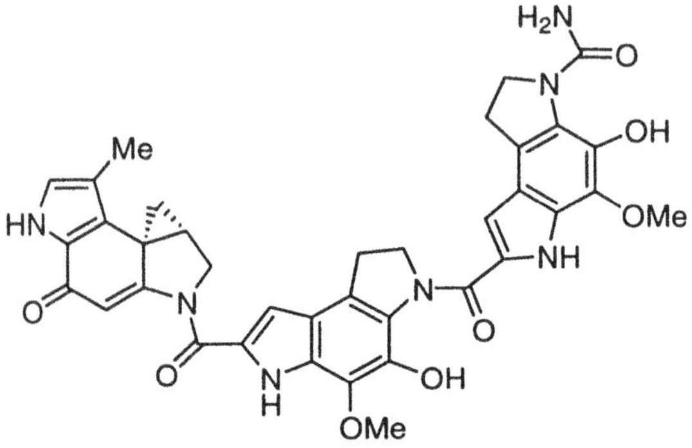

G. Bifulco, J.A. Smith, W.J. Chazin, L. Gomez-Paloma

5, (+)-CC-1065

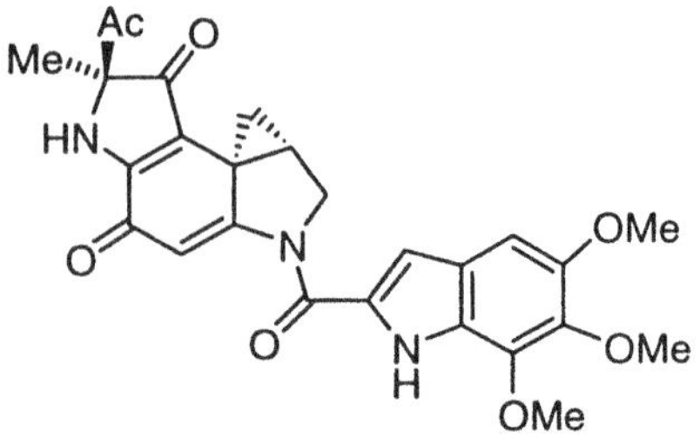

6, (+)-Duocamycin A (DA)

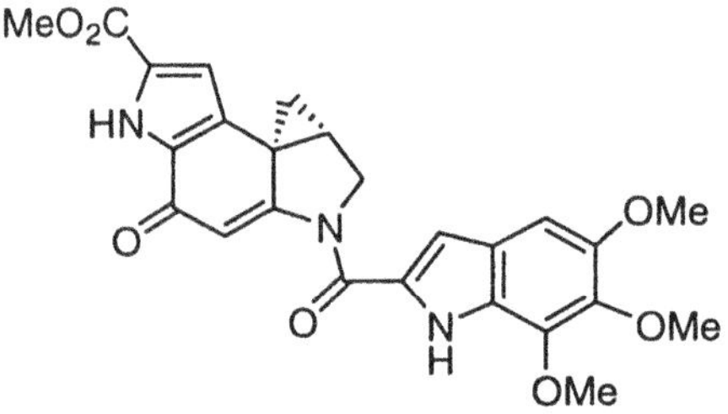

7, (+)-Duocamycin SA (DSA)

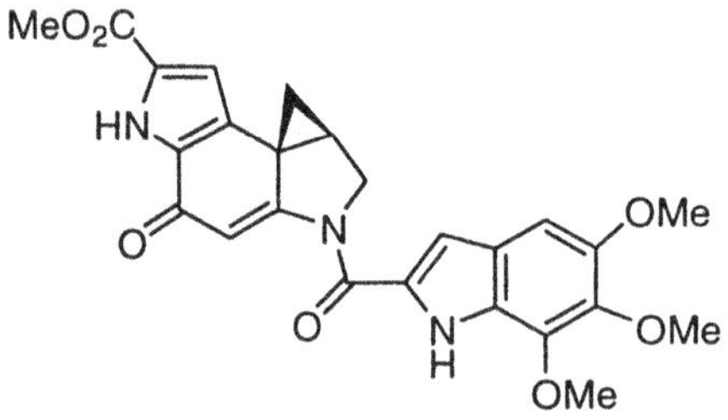

8, (-)-DSA

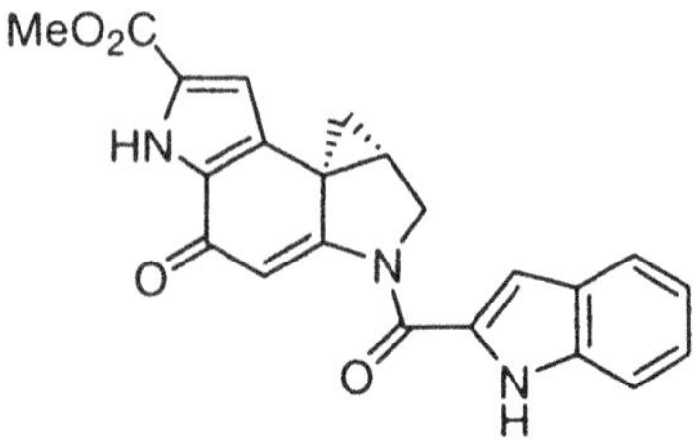

9, (+)-DSI

Fig. 9. Chemical structures of some natural and synthetic derivatives of the duocarmycin family.

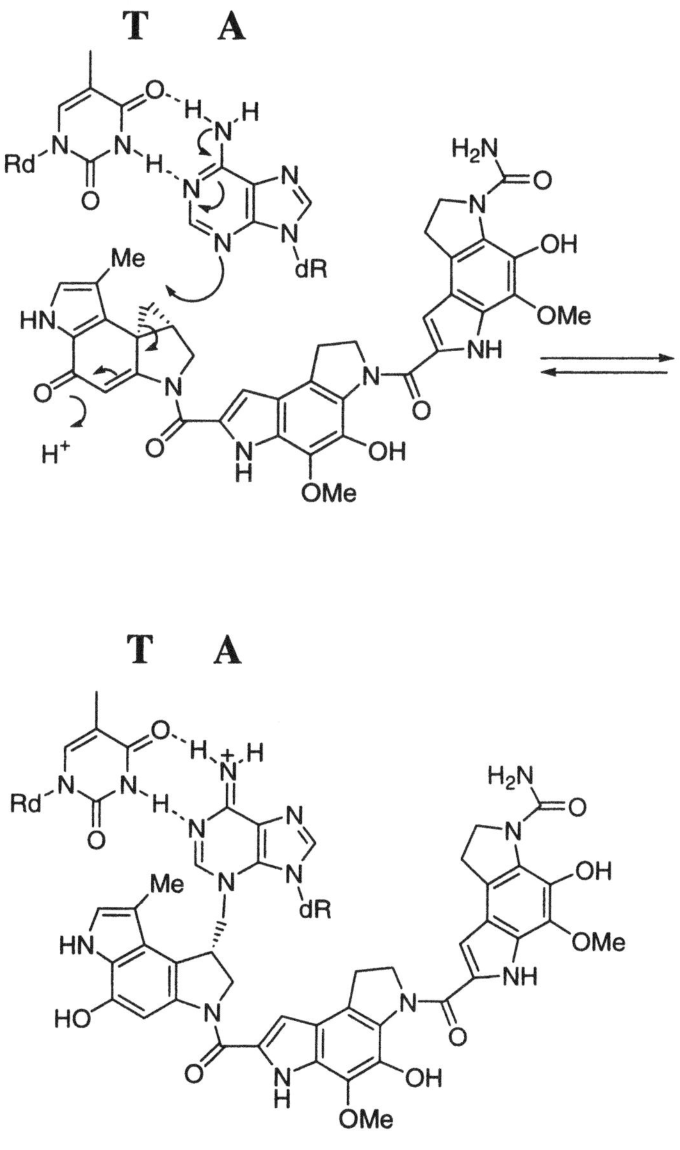

Fig. 10. Mechanism of alkylation of duplex DNA by CC-1065. The most important features of the *alkylation site* model and the *noncovalent binding* model are shown.

Studies on the mechanism of the alkylation reaction have established a link between the change of conformation induced by the binding of the agent to DNA and the catalysis of the alkylation reaction [26]. Structural information obtained through NMR studies of DNA complexes of three duocarmycin derivatives, the natural and unnatural enantiomers of DSA (**8**) [39, 40], and the indole derivative DSI (**9**) [41], have provided critical evidence establishing the molecular basis for the chemical, functional and biological properties of duocarmycins and CC-1065.

B. Sequence selectivity of duocarmycins and CC-1065

As outlined in the introduction, CC-1065 and duocarmycins are reminiscent of other minor groove binding ligands such as distamycin A and its derivatives. These flat aromatic molecules have a crescent shape and show a marked binding preference (at least in the 1:1 complex) towards AT-rich DNA sequences. The origin of this sequence selectivity has been explained on the basis of two main observations that come from crystallographic and NMR studies:

a) the minor groove of AT-rich tracts is deep and narrow and sterically accessible, as opposed to that observed in GC-rich regions where the minor groove is wider, shallow and hindered by the guanine amino group that protrudes towards the exterior of the groove;

b) the main stabilizing interactions that can be invoked to explain sequence selectivity of this type of ligands are hydrogen bonding, Van der Waals forces and hydrophobicity.

Therefore, flat molecules capable of adopting a partially curved shape, such as CC-1065 and duocarmycins should prefer AT-rich regions of DNA because the absence of the guanine amino group allows a deeper penetration of the ligand into the minor groove. In addition, the shorter distance between the walls of the groove in AT-rich regions allows optimal Van der Waals contacts with the aromatic systems of the ligand leading to a favorable non-covalent binding stabilization.

A wealth of information has been collected in recent years on the details of the binding preference of duocarmycins and other related molecules. The data indicate that when the *binding subunit* is absent, like in N-BOC-DSA, the DNA alkylation pattern does not show any significant degree of selectivity, in accordance with the hypothesis that the selectivity of alkylation is governed by the noncovalent recognition deriving from the right-hand subunit.

The binding activation hypothesis is completely general and equally valid for all pairs of enantiomers. However, a given AT-rich site may be highly favorable for alkylation by one enantiomer but not the other (sequence selectivity). This is of course the consequence of the different spatial positioning of the cyclopropyl ring in the two enantiomers. For further clarification, the alkylation of an adenine residue in a given DNA duplex requires that the two enantiomers must bind and extend in the minor groove in the opposite direction one from another (reversed binding orientation) in order for the cyclopropyl ring to be positioned for nucleophilic attack on this residue. Another possibility is that alkylation will occur on opposite strands with both enantiomers binding in the same orientation within the binding site, again as a consequence of the different spatial arrangement of the cyclopropyl ring.

Important insights into the structural origin of the sequence-selectivity were derived from the examination of the characteristics of the noncovalent binding to DNA of a series of compounds containing only *binding subunits*, such as 10–14 [41]. Plasmid DNA protection assays revealed preference for AT-rich DNA tracts, but also implicated there is an optimal length for these oligomers. Indeed, compounds with three CPI subunits (12) display the best binding properties, spanning five base pairs or half a helical turn. This appears to constitute the largest accessible site for synchronous binding of both ends of the agent. Higher order oligomers (e.g. 13 and 14 with four or five subunits) bind to DNA less tightly presumably because they fall out of the register and further interactions in the DNA minor groove are not highly productive. The compound (12) with three subunits can be thought as the noncovalent functional analog of CC-1065. These studies also documented the small importance of the hydroxy and methoxy group present in the CC-1065 structure on the binding affinity and sequence-selectivity.

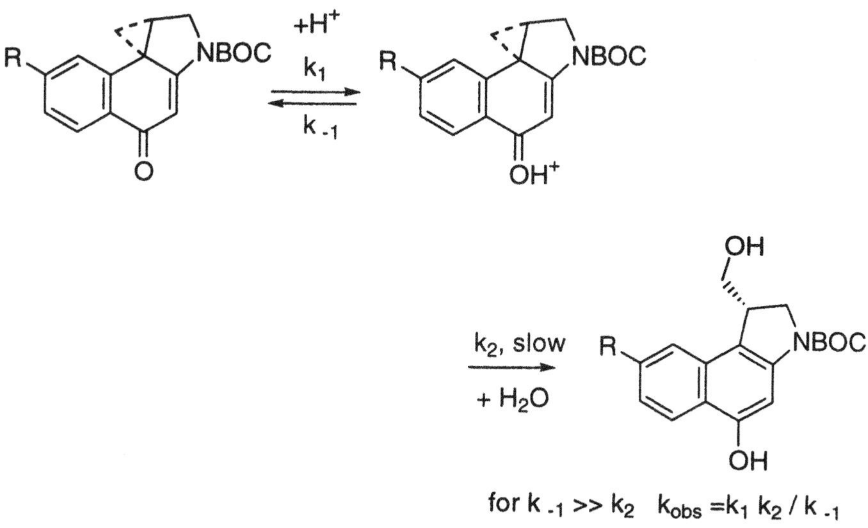

Fig. 11. Binding parameters of the solvolysis reaction of a DSA modified alkylation subunit.

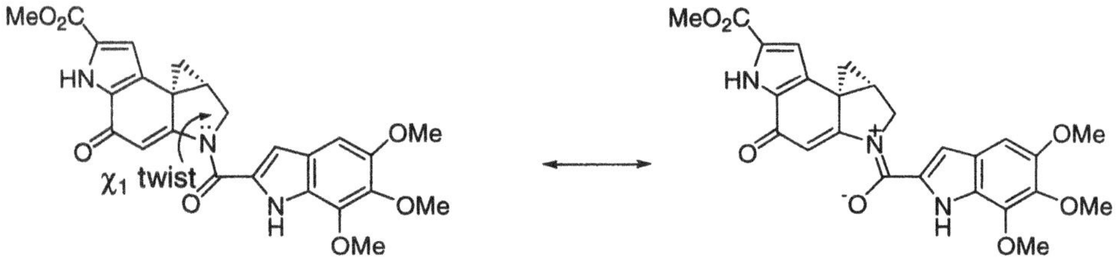

- DNA bound agent maintains full amide
- Vinylogous amide stabilization diminished (χ_1 twist)
- Cyclohexadienone structure destabilized

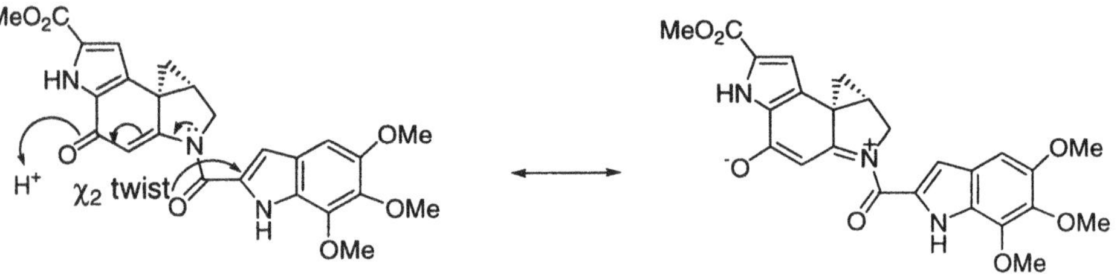

- DNA bound agent contains twisted amide (χ_2 twist)
- Twisted amide releases electron density into vinilogous amide
- Results in more efficient C4 carbonyl protonation

Fig. 12. Implications for DNA alkylation associated with the twisting of χ_1 and χ_2 torsion angles upon binding to duplex DNA.

C. Mechanism of alkylation of duplex DNA: in situ activation of the ligand

Duocarmycins, despite the presence of a cyclopropane ring, are relatively unreactive species when exposed to small nucleophiles [42] in acidic media and are indefinitely stable at pH = 7 where they exhibit no measurable rate of reaction with nucleophiles. For instance, duocarmycin SA (**7**) has a $t_{1/2}$ of 177 h at pH = 3. The unusual stability of these molecules is generally explained in terms of ground state stabilization due to the high degree of conjugation of the alkylating subunit.

The alkylation reaction in the minor groove of duplex DNA containing high-affinity sites is, in contrast, faster by a factor of $10^{10}-10^{12}$, proceeding smoothly at pH = 7.4 in a few hours even at 4°C under pseudo first-order reaction conditions (i.e. in presence of excess DNA duplex) [43].

Several proposals have been advanced to explain this remarkable catalytic phenomenon. One hypothesis attributes the sequence dependent acceleration of the rate of alkylation reaction to differential reactivities of the adenine N3 nucleophiles associated with the different DNA sequences [44]. A second model assigns great importance to the C4 carbonyl protonation and assumes that this occurs via the DNA phosphate backbone in a sequence dependent manner [45]. A further proposal invokes a conformationally trapped DNA reactivity associated with bent DNA, again in a sequence dependent fashion [46]. The model based on the C4 carbonyl protonation is attractive because it is consistent with the solvolysis of duocarmycins observed in acidic media. However, consideration that the phosphate backbone is fully ionized at pH = 7.4 (with 0.0001–0.00004% of sites protonated) and that the reaction in the absence of DNA is quite slow even at pH = 5 suggests that this explanation is not correct, even taking into account the greater local concentration of protons in the proximity of the anionic phosphate groups. Electrophilic activation at the C4 carbonyl site could be more readily explained by complexation with one of the cations that surround the DNA backbone.

There is now considerable experimental evidence indicating that alkylation selectivity is controlled by binding selectivity, and that the source of catalysis of the DNA alkylation reaction is a DNA induced conformational change of the agent in the bound state [47]. In this view, the rate acceleration can be explained in terms of ground state destabilization associated with the disruption of the extensive conjugation of the alkylating subunit upon binding to DNA. This destabilization of the substrate, which results from the twisting of the linkage between N2 and the cyclohexadienone, would destroy the vinylogous amide stabilization responsible for the unusual stability of the duocarmycin agents, thereby increasing the inherent reactivity towards nucleophiles. This activation is not alkylation site dependent but, in contrast, is a general consequence of the helical conformation that the agent adopts when bound within the minor groove of duplex DNA.

The electronic structure of duocarmycin (as well as that of CC-1065) can be described in terms of cross-conjugation of the N2 nitrogen atom with two other π systems, i.e. the N2 lone pair is conjugated to both the cyclohexadienone (left-hand) and the amide carbonyl (right-hand). When the agent is bound, there are two primary ways that the two subunits of duocarmycin can be twisted to follow the DNA minor groove.

One possibility is for the N2 lone pair to remain fully conjugated to the cyclohexadienone, with the diminished amide resonance. This would pump electron density into

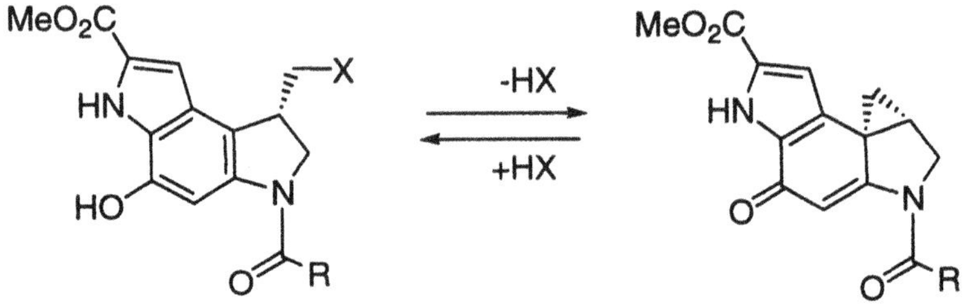

Fig. 13. Reversibility of the DNA alkylation reaction.

the vinylogous amide leading to an increased basicity of the C4 carbonyl and/or more effective cation complexation. Alternatively, the N2 lone pair remains conjugated to the amide carbonyl and the cyclohexadienone loses the benefit of vinylogous amide stabilization. This would provide a much more reactive intermediate prone to nucleophilic addition even without the intervention of a proton or cation complexation to the C4 carbonyl. Although there is no evidence obtained directly on the transition state, an examination of the reaction product would show a significant twist in the χ_2 angle if full vinylogous amide conjugation were more important, whereas full amide conjugation would lead to more severe twist of the χ_1 angle. Indeed, high resolution structural analysis of duocarmycin-DNA complexes has shown that the majority of the inter-subunit twist is found in χ_1. For example in the complex of (+)-DSA with a defined high affinity DNA duplex, one half of the 44° twist is found in χ_1 [32].

It is noteworthy that this *in situ* activation mechanism for DNA alkylation by duocarmycin can be compared to an enzymatic process where the DNA plays the role of the enzyme. There are many ingredients of typical enzymatic reactions, such as distortion of crucial functional groups to decrease activation energy, close proximity of the reactive species (the nucleophile) to the substrate, and possibly site directed protonation or metal complexation. This observation becomes even more intriguing when considering the common belief that, among nucleic acids, only RNA possesses catalytic activity. Interestingly, from the kinetic data reported to date, it is possible to fit to the Michaelis-Menten equation with a $K_M = 0.4$ mM [48].

One further consideration relevant to DNA alkylation is that cyclopropane ring opening induced by DNA nucleophiles is intrinsically reversible [49]. Indeed it is surprisingly easy to introduce the cyclopropane ring functionality through an Ar-3' spirocyclization. So, the question of why alkylated DNA possesses such unusual stability cannot be answered in terms of irreversibility of the 3-membered ring opening reaction. Indeed, the stability of duocarmycin-DNA adducts has been attributed to non-covalent binding stabilization provided by the right-hand subunit [50]. This conclusion is supported by the observation that agents characterized by a larger size of the binding subunit(s), e.g. CC-1065 (5), display an increased stability when bound to DNA and conversely agents possessing only

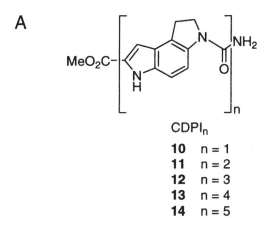

CDPI$_n$

10 n = 1
11 n = 2
12 n = 3
13 n = 4
14 n = 5

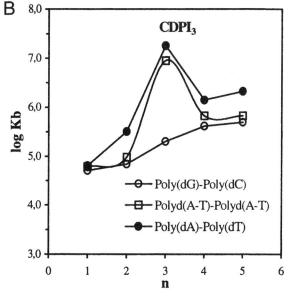

Fig. 14. (A) Chemical structures of some oligomers (n = 1–5) of the DSA noncovalent subunit; (B) Plot of the binding affinity (expressed as log K_b) of these oligomers to different DNA duplexes as a function of the number of subunits n.

the alkylating subunit (such as N-Boc-DSA) exhibit more facile retro-alkylation. The ground state destabilization due to disruption of the vinylogous amide, also provides an explanation for the stability of the product, since the DNA adduct will not be destabilized by the twisting of the agent in the DNA minor groove.

D. The structural basis for in situ activation of DNA alkylation

A wealth of information on the structural basis for *in situ* activation of DNA alkylation has been obtained through three high resolution NMR structures of 1:1complexes between (+) DSA, (−)-DSA and (+)-DSI (**7–9**) bound to d(GACTAATTGAC)·(GTCAAT-TAGTC) containing the high affinity AATTA site [39–41].

As pointed out in the preceding sections, the rigid aromatic subunits of the duocarmycins are coplanar when the agents are in the free state, due to the extended conjugation. Upon binding in the DNA minor groove, the two subunits twist with respect to each other as they are driven to maximize hydrophobic contacts with the DNA (Fig. 15). The twist of the two subunits is distributed across the torsion angles $\chi 1$, $\chi 2$, and $\chi 3$ in the linkage between them (Fig. 12). The sum of the $\chi 1$, $\chi 2$, and $\chi 3$ angles has been defined as the overall inter-subunit twist angle. The (−)-DSA adduct is found to have a relatively small binding-induced twist angle between the two subunits of the ligand relative to the two (+) adducts ($21° \pm 3°$ for (−)-DSA, $44° \pm 3°$ for (+)-DSA, and $37° \pm 2°$ for (+)-DSI). The inter-subunit twist angle is best viewed down the minor groove of the complex, as shown in Figure 16. The inherent local flexibility of the DNA enables adjustment of the conformation differently in response to alkylation at different sites. However, the handedness of the DNA molecule itself imposes an important constraint on the (+) agent that is not present for the (−) agent. Since DNA has a left-handed helical twist, the natural tendency is for the trimethoxyindole subunit of DNA to be forced to twist in a counter-clockwise fashion relative to the alkylation subunit when bound in the minor groove.

The (+)-DSI derivative (**5**) has been observed to penetrate more deeply into the DNA minor groove than the (+)-DSA parent compound due to the lack of the bulky methoxy group at C24. This correlates with the smaller degree of inter-subunit twist in (+)-DSI relative to (+)-DSA ($37°$ vs. $44°$). In the (+)-DSI adduct, the unhindered indole subunit slides closer to the floor of the minor groove, with the edge of the indole itself extending into the cleft between A16 and T17 that is occupied by the C24 methoxy group in both DSA adducts. Thus, the nature of the structural adaptation is clearly different than what is observed for (−)-DSA, even though the net effect is a lower value of the overall inter-subunit twist angle in both the (+)-DSI and (−)-DSA adducts.

Among the three component torsions contributing to inter-subunit twist, $\chi 1$ exhibits the largest perturbation from planarity in all three adducts, consistent with the expectation that the energy barrier for rotation about $\chi 1$ is the smallest of the three. Demonstration of the excellent correlation between binding-induced twist at $\chi 1$ and the relative efficiency of DNA alkylation supports the binding-induced activation hypothesis. The alkylation efficiency of (+)-DSA is considerably greater than that of either (−)-DSA or (+)-DSI, which are roughly equal to one another. In detail, the ratio is 30:2:1, with (+)-DSA being the most efficient alkylator, and (+)-DSI being the least. This correlates well with the values measured for $\chi 1$: (+)-DSA yields the largest $\chi 1$ value ($22.4° \pm 0.5°$), consistent with its considerably higher alkylation efficiency compared with either of the other two agents. Taking experimental error into account, (−)-DSA and (+)-DSI yield values that are nearly the same ($11.2° \pm 1.2°$ and $14.3° \pm 0.3°$, respectively), consistent with their similar alkylation efficiencies.

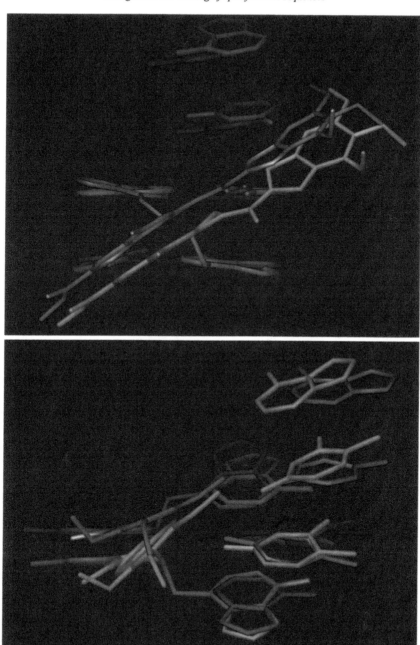

Fig. 15. Comparison of the (+)- and (−)-DSA adducts (orange and blue, respectively) to d-(GACTAATTGAC)·
d-(GTCAATTAGTC). The upper panel depicts the difference in the positioning of the ligand within the DNA
minor groove. The lower panel highlights the corresponding difference in the ligand inter-subunit twist angle.

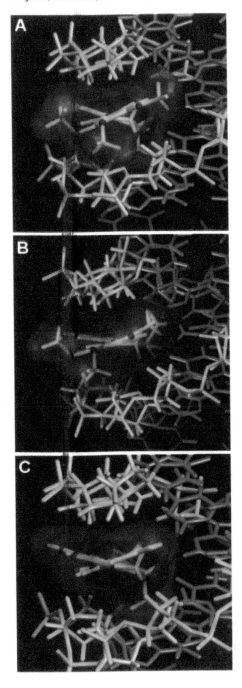

Fig. 16. Comparison of the inter-subunit twist of the (A) (+)-DSA, (B) (–)-DSA, and (C) (+)-DSI adducts to d-(GACTAATTGAC)·d-(GTCAATTAGTC). The two strands of DNA are depicted in pink and cyan. The ligands are depicted in white with heteroatoms as colored spheres, and include molecular surfaces in blue.

It is interesting to note that even though (−)-DSA and (+)-DSI exhibit comparable values for $\chi 1$, their overall twist angles ($\chi 1 + \chi 2 + \chi 3$) are significantly different (21° and 37°, respectively). This is due to differences in the partitioning of the overall twist between the three χ angles. The two DSA enantiomers have identical $\chi 1 : \chi 2 : \chi 3$ ratios of 2:1:1 ($\chi 1 > \chi 2 = \chi 3$), while (+)-DSI is distinct with $\chi 1 \cong \chi 2 > \chi 3$. These results show that although the overall molecular mechanism of interaction involves a binding-induced twist of one subunit relative to another, it is the perturbation of $\chi 1$ that most directly relates to DNA alkylation reactivity.

Acknowledgements

We thank the American Cancer Society for past support of ligand-DNA research in the Chazin laboratory, and acknowledge the current support and interest of the Vanderbilt Center for Molecular Toxicology (NIEHS P30 ES00267) and the Vanderbilt-Ingram Cancer Center.

References

[1] Reviewed in: K.C. Nicolaou, W.-M. Dai, *Angew. Chem. Int. Ed. Engl.* 30 (1991) 1387–1416; K.C. Nicolaou, A.L. Smith, *Acc. Chem. Res.* 25 (1992) 497–503; M.E. Lee, G.A. Ellestad, D.B. Borders, *Acc. Chem. Res.* 25 (1992) 235–243; K.C. Nicolaou, A.L. Smith, E.W. Yue, *Proc. Natl. Acad. Sci. U.S.A.* 90 (1993) 5881–5888.

[2] K.C. Nicolaou, P. Stabila, B. Esmaeli-Azad, W. Wrasidlo, A. Hiatt, *Proc. Natl. Acad. Sci., U.S.A.* 90 (1993) 3142–3146.

[3] G.A. Ellestad, P.R. Hamann, N. Zein, G.O. Morton, M.M. Siegel, M. Pastel, D.B. Borders, W.J. McGahren, *Tetrahedron Lett.* 30 (1989) 3033.

[4] N. Zein, M. Poncin, R. Nilakantan, G.A. Ellestad, *Science* 244 (1989) 697.

[5] W.D. Ding, G.A. Ellestad, *J. Am. Chem. Soc.* 113 (1991) 6617–20.

[6] K.C. Nicolaou, S.-C. Tsay, T. Suzuki, G.F. Joyce, *J. Am. Chem. Soc.* 114 (1992) 7555–7557.

[7] L. Gomez Paloma, J.A. Smith, W.J. Chazin, K.C. Nicolaou, *J.Am.Chem.Soc.* 116 (1994) 3697–3708; J.A. Smith, L. Gomez -Paloma, D.A. Case, W.J. Chazin, *Magn. Reson. Chem.* 34 (1996) S147–S155.

[8] T. Li, Z. Zeng, V.A. Estevez, K.U. Baldenius, K.C. Nicolaou, G.F. Joyce, *J.Am.Chem.Soc.* 116 (1994) 3709–3715.

[9] R.C. Hawley, L.L. Kiessling, S.L. Schreiber, *Proc. Natl. Acad. Sci.* 86 (1989) 1105–1109.

[10] C. Bailly, M.J. Waring, *J. Am. Chem. Soc.* 117 (1995) 7311–7319.

[11] S. Walker, J. Murnick, D. Kahne, *J. Am. Chem. Soc.* 115 (1993) 7954–7961.

[12] S. Walker, A.H. Andreotti, D. Kahne, *Tetrahedron* 50 (1994) 1351–1360.

[13] S. Walker, D. Gange, V. Gupta, D. Kahne, *J. Am. Chem. Soc.* 116 (1994) 197–3206.

[14] M. Mrksich, P.B. Dervan, 115 (1993) 2572–2576.

[15] J.W. Lown, K. Krowicki, U.G. Bhat, A. Skorobogaty, B. Ward, J.C. Dabrowiak, *Biochemistry* 25 (1986) 7408–7416.

[16] K.C. Nicolaou, K. Ajito, H. Komatsu, B.M. Smith, T. Li, M.G. Egan, L. Gomez- Paloma, *Angew. Chem. Int. Ed.* 34 (1995) 576–578.

[17] K.C. Nicolaou, B.M. Smith, K. Ajito, H. Komatsu, L. Gomez-Paloma, Y. Tor, *J. Am. Chem. Soc.* 118 (1996) 2303–2304.

[18] C. Liu, B.M. Smith, K. Ajito, H. Komatsu, L. Gomez-Paloma, T. Li, E.A. Theodorakis, K.C. Nicolaou, P.K. Vogt, *Proc. Natl. Acad. Sci.* 93 (1996) 940–944.

[19] G. Bifulco, A. Galeone, L. Gomez-Paloma, K.C. Nicolaou, W.J. Chazin, *J. Am. Chem. Soc.* 119 (1996) 8817–8824.

[20] G. Bifulco, A. Galeone, K.C. Nicolaou, W.J. Chazin, L. Gomez-Paloma, *J. Am. Chem. Soc.* 120 (1998) 7183–7191.

[21] D.L. Boger, D.S. Johnson, *Angew. Chem. Int. Ed. Engl.* 35 (1996) 1438–1474.

[22] D.L. Boger, D.S. Johnson, W. Wrasidlo, *Febs. Lett.* 4 (1994) 631–636.

[23] D.L. Walker, J.M. Reid, M.M. Ames, *Cancer Chemother. Pharmacol.* 34 (1994) 317–322.

[24] E. Kobayashi, A. Okamoto, M. Asada, M. Okabe, S. Nagamura, A. Asai, H. Saito, K. Gomi, T. Hirata, *Cancer Res.* 54 (1994) 2404–2410.

[25] D.L. Boger, H. Zarrinmayeh, S.A. Munk, P.A. Kitos, O. Suntornwat, *Proc. Natl. Acad. Sci. U.S.A.* 88 (1991) 1431–1435.

[26] D.L. Boger, C.W. Boyce, R.M. Garbaccio, J.A. Goldberg, *Chem. Rev.* 97 (1997) 787–828.

[27] D.L. Boger, R.M. Garbaccio, *Bioorganic & Medicinal Chemistry* 5 (1997) 263–276.

[28] P.A. Aristoff, J.P. McGovren, *Drug News Perspect* 6 (1993) 229–234; R.D. Hightower, B.U. Sevin, J. Perras, H. Nguyen, R. Angioli, M. Untch, H. Averette, *Cancer Invest.* 11 (1993) 276–282; D.L. Walker, J.M. Reid, M.M. Ames, *Cancer Chemoter. Pharmacol.* 34 (1994) 317–322; E. Kobayashi, A. Okamoto, M. Asada, M. Okabe, S. Nagamura, A. Asai, H. Saito, K. Gomi, T. Hirata, *Cancer Res.* 54 (1994) 2404–2410.

[29] D.E. Nettleton, J.A. Bush, W.T Bradner, US Patent 4 301 248, *Chem. Abstr.* 96 (1982) 33362e.

[30] D.G. Martin, C.G. Chidester, D.J. Duchamp, S.A. Mizsak, *J. Antibiot.* 33 (1980) 902–903; C.G. Chidester, W.C. Krueger, S.A. Mizsak, D.J. Duchamp, D.G. Martin, *J. Am.Chem. Soc.* 103 (1981) 7629–7635.

[31] B.K. Bhuyan, K.A. Newell, S.L. Crampton, Von D.D. Hoff, *Cancer Res.* 42 (1982) 3532–3537.

[32] M. Ichimura, T. Ogawa, K. Takahashi, E. Kobayashi, I. Kawamoto, T. Yasuzawa, I. Takahashi, H. Nakano, *J. Antibiot.* 43 (1990) 1037–1038.

[33] T. Yasuzawa, Y. Saitoh, M. Ichimura, I. Takahashi, H. Sano, *J. Antibiot.* 44 (1991) 445–447; M. Ichimura, T. Ogawa, S. Katsumata, K. Takahashi, I. Takahashi, H. Nakano, *J. Antibiot.* 44 (1991) 1045–1053.

[34] M. Ichimura, T. Ogawa, K. Takahashi, A. Mihara, I. Takahashi, H. Nakano, *Oncol. Res.* 5 (1993) 165–171; J.P. McGroven, G.L. Clarke, E.A. Pratt, T.F. DeKoning, *J. Antiobiot.* 37 (1984) 63–70.

[35] Reviewed in: D.L. Boger, D.S. Johnson, *Angew. Chem. Int. Ed. Engl.* 35 (1996) 1439; D.L. Boger, D.S. Johnson, *Proc. Natl. Acad. Sci., U.S.A.* 92 (1995) 3642; D.L. Boger, *Acc. Chem. Res.* 28 (1995) 20; D.L. Boger, Advances in heterocyclic natural product synthesis. In: W.H. Pearson, ed., vol. 2 (p. 1). JAI, Greenwich, 1992; L.H. Hurley, P.H. Draves, Molecular aspects of anticancer drug-DNA interactions. In: S. Neidle, M. Waring, eds, vol. 1 (p. 89). CRC, Ann Arbor, 1993; M.A. Warpehoski, Advances in DNA sequence specific agents. In: L.H. Hurley, ed., vol. 1 (p. 217). JAI, Greenwich, 1992; P.A. Aristoff, Advances in medicinal chemistry. In: B.E. Maryanoff, C.E. Maryanoff, eds, vol. 2 (p. 67). JAI, Greenwich, 1993; M.A. Warpehoski, J.P. McGovren, M.A. Mitchell, L.H. Hurley, Molecular basis of specificity in nucleic acid-drug interactions. In: B. Pullman, J. Jortner, eds (p. 531). Kluwer, The Netherlands, 1990; M.A. Warpehoski, L.H. Hurley, *Chem. Res. Toxicol.* 1 (1998) 315.

[36] M.A. Warpehoski, I. Gebhard, R.C. Kelly, W.C. Krueger, L.H. Li, J.P. McGovren, M.D. Prairie, N. Wicnienski, W. Wierenga, *J. Med. Chem.* 31 (1988) 590–603.

[37] D.L. Boger, W. Yun. *J. Am. Chem. Soc.* 116 (1994) 5523–5524.

[38] P.A. Aristoff, J.P. McGovren, *Drug News Perspect.* 6 (1993) 229–234.

[39] P.S. Eis, J.A. Smith, J.M. Rydzewski, D.A. Case, D.L. Boger, W.J. Chazin, *J. Mol. Biol.* 272 (1997) 237–252.

[40] J.A. Smith, G. Bifulco, D.A. Case, D.L. Boger, L. Gomez-Paloma, W.J. Chazin, *J. Mol. Biol.* 300 (2000) 1195–1204.

[41] J.R. Schnell, R.R. Ketchem, D.L. Boger, W.J. Chazin, *J. Am. Chem. Soc.* 121 (1999) 5645–5652.

[42] D.L. Boger, B.J. Invergo, R.S. Coleman, H. Zarrinmayeh, P.A. Kitos, S.C. Thompson, T. Leong, L.W. McLaughlin, *Chem. Biol. Interact.* 73 (1990) 29–52; D.L. Boger, S.M. Sakya, *J. Org. Chem.* 57 (1992) 1277–1284.

[43] D.L. Boger, T. Ishizaki, P.A. Kitos, O. Suntorwat, *J. Org. Chem.* 55 (1990) 5823.

[44] H. Sugiyama, M. Hosoda, I. Saito, A. Asai, H. Saito, *Tetrahedron Lett.* 31 (1990) 7197.

[45] M.A. Warpehoski, L.H. Hurley, *Chem. Res. Toxicol.* 1 (1988) 315.

[46] C.H. Lin, J.M. Beale, L.H. Hurley, *Biochemistry.* 30 (1991) 3597; L.H. Hurley, M.A. Warpehoski, C.S. Lee, J.P. McGroven, T.A. Scahill, R.C. Kelly, M.A. Mitchell, N.A. Wicnineski, I. Gebhard, P.D. Johnson, V.S. Bradford, *J. Am. Chem. Soc.* 112 (1990) 4633; D.G. Martin, R.C. Kelly, W. Watt, N. Wicnienski, S.A. Mizsak, J.W. Nielsen, M.D. Prairie, *J. Org. Chem.* 53 (1988) 4610.

[47] C.H. Lin, D. Sun, L.H. Hurley, *Chem. Res. Toxicol.* 4 (1991) 21; C.S. Lee, D. Sun, R. Kizu, L.H. Hurley, *Chem. Res. Toxicol.* 4 (1991) 203; C.H. Lin, G.C. Hill, L.H. Hurley, *Chem. Res. Toxicol.* 5 (1992) 167; Z.M. Ding, R.M. Harshey, L.H. Hurley, *Nucl. Acids Res.* 21 (1993) 4281; D. Sun, C.H. Lin, L.H. Hurley, *Biochemistry* 32 (1993) 4487; A.S. Thompson, D. Sun, L.H. Hurley, *J. Am. Chem. Soc.* 117 (1995) 2371.

[48] D.L. Boger, J. Zhou, H. Cai, *Bioorg. Med. Chem.* 4 (1996) 859; D.L. Boger, D.S. Johnson, *J. Am. Chem. Soc.* 117 (1995) 1443.

[49] M.A. Warpehoski, D.E. Harper, *J. Am. Chem. Soc.* 117 (1995) 2951.

[50] D.L. Boger, D.S. Johnson, W. Yun, *J. Am. Chem. Soc.* 116 (1994) 1635–1656; D.L. Boger, W. Yun, *J. Am. Chem. Soc.* 115 (1993) 9872–9873.

[51] D.L. Boger, H. Zarrinmayeh, S.A. Munk, P.A. Kitos, O. Suntornwat, *Proc. Natl. Acad. Sci., U.S.A.* 88 (1991) 1431–1435.

After preparation of this manuscript, important new studies appeared indicating that the indirect readout mechanism must play a role in the binding of calicheamicin aryltertrasaccharides. A. Kalben, S. Pal, A. Santona, A. Hamilton, S. Walker, D. Gange, K. Biswas, D. Kahne. *J. Am. Chem. Soc.* 122 (2000) 8403–8412; K. Biswas, S. Pal, J.D. Carbeck, D. Kahne. *J. Am. Chem. Soc.* 122 (2000) 8413–8420; A.A. Salzberg, P.C. Dedon. *Biochem.* 39 (2000) 7605–7612.

Advances in DNA Sequence-specific Agents 04 (2002) 75–103

Enediyne antibiotic neocarzinostatin as a radical-based probe of bulged structures in nucleic acids

Zhen Xi and Irving H. Goldberg*

Department of Biological Chemistry and Molecular Pharmacology, Harvard Medical School, Boston, MA 02115

1. Introduction

A. Enediyne antibiotics are efficient DNA duplex cleavers

Enediyne antitumor antibiotics represent a novel family of chemical compounds with extraordinary biological and chemical properties [1–3]. It has been and remains an active area of research since the discovery of a new antibiotic "protein" neocarzinostatin from *Streptomyces carzinostaticus* var. F-41 in 1965 [4,5]. This agent was found to be capable of inhibiting DNA synthesis and inducing DNA degradation in cells [6]. Efforts to understand the biochemical behavior of this agent led to the finding that its actual DNA-damaging activity comes from a previously unidentified non-protein chromophore (NCS-chrom) [7–11] through a free-radical mechanism involving hydrogen atom abstraction from the sugar residues of DNA [12]. The observed biological and pharmacological properties place the enediyne antibiotics among the most potent antitumor agents ever found [2] The clinical application of neocarzinostatin against acute leukemia and certain solid tumors (stomach, colon, kidney, and bladder) in Japan and the phase III clinical trial of antibody-conjugated calicheamicin against acute myelogenous leukemia in the United States highlight the medicinal potential of this class of antibiotics.

Neocarzinostatin (NCS) has been subjected to extensive mechanistic study. It is generally accepted that the non-protein chromophore (NCS-chrom) of neocarzinostatin is responsible for its DNA cleavage activity and that apo-neocarzinostatin (the protein part of the chromoprotein) protects NCS-chrom from degradation and delivers the chromophore to the targeted DNA *in vivo* [13].

*Corresponding author
E-mail address: Irving_Goldberg@hms.harvard.edu
Advances in DNA Sequence-specific Agents, Volume 4 G. B. Jones (Editor)

Early studies on *in vitro* DNA scission revealed that thiol-activated NCS-chrom produced primarily single-stranded (SS) breaks with base selectivity. About 75% of breaks were at T residues (T > A >> C > G) [13]. Whereas SS lesions occur mainly as breaks due to 5′-chemistry or to a much lesser degree as abasic sites (and breaks) due to 4′-chemistry, double-stranded (DS) lesions (DS break or abasic site with a closely opposed SS break) involve a mixture of chemistry (5′ and 1′ or 4′) at a staggered lesion site (two residues apart). DS lesions are sequence specific (either AG$\underline{C}$·G$\underline{T}$ or AG$\underline{T}$·A$\underline{T}$) (attack sites underlined). DS lesions at AG$\underline{C}$·G$\underline{T}$ sites consist of an apyrimidinic abasic site due to 2-deoxyribonolactone formation (1′-chemistry) at the $\underline{C}$ residue of AG$\underline{C}$ and a break due to 5′-nucleoside aldehyde formation (5′-chemistry) at the $\underline{T}$ residue two nucleotides to the 3′ side on the complementary strand. Detailed analysis of the DS lesions induced by glutathione-activated NCS-chrom at a model AG$\underline{T}$·A$\underline{T}$ in the AP-1 transcription factor-binding site showed that 89% of the DS lesions at the $\underline{T}$ of AG$\underline{T}$ were due to C-4′ hydrogen abstraction and 11% from C-5′ hydrogen abstraction [14]. Seventy-four per cent of the 4′-chemistry was in the form of a 4′-hydroxylated abasic lesion and the remainder in a 3′-phosphoglycolate-ended fragment. The break at the $\underline{T}$ of A$\underline{T}$ on the complementary strand was more than 90% due to 5′-chemistry, generating mainly 5′-nucleoside aldehyde. Deuterium abstraction experiments [15,16], molecular modeling studies [17], and NMR solution structure [18,19] of the complex formed between the glutathione-postactivated drug adduct and an AG$\underline{C}$·G$\underline{T}$ site-containing oligonucleotide indicate that the DS lesions result from the concerted action of a single drug on both strands. The radical center at C2 of the drug is involved in attack at the $\underline{C}$ of AG$\underline{C}$ (and the $\underline{T}$ of AG$\underline{T}$), while that at C6 abstracts hydrogen from the $\underline{T}$ two bases to the 3′ side on the complementary strand. The C6 radical is involved in both SS and DS lesions, whereas that at C2 is only involved in DS lesions. Deuteration and point mutations in DS lesion sites have been found to dramatically alter the chemistry of DNA damage [20–22]. These findings provide support for the role of minor groove microstructure in determining the chemical mechanism of DNA damage and underscore the usefulness of NCS-chrom as a dynamic probe of DNA microheterogeneity.

Thiol-activated NCS-chrom has also been found to generate staggered (two nucleotides in the 3′-direction) DS lesions in DNA·RNA hybrids, involving 5′-chemistry on the DNA strand and 1′-chemistry on the RNA strand [23]. The strong deuterium isotope effect on H1′ abstraction at the damaged uridylate confirmed that the abasic site formation was due to 1′-chemistry.

B. Nucleic acid bulge is an important biological motif

Bulge structures in nucleic acids, involving one or more unpaired bases, are of general biological significance [24,25]. Bulges are very common in RNA, where they play important roles in protein binding recognition [26]. On the other hand, DNA bulges are believed to be involved in frameshift mutation and are the products of imperfect homologous recombination or slipped mispairing during the replication of DNA. [27–29]. DNA bulges have been shown to be targets for DNA repair enzymes [30,31]. Multi-base bulges, such as triplet repeats, have been implicated in their unstable expansion during

DNA replication in inherited neurodegenerative diseases like Huntington's disease [32]. Consequently, nucleic acid bulges have been the subject of intense study. The local conformations of single-base bulges in several oligodeoxynucleotides have been characterized by NMR spectroscopy [33,34]. It has been found that the equilibrium between a bulge nucleobase stacking in or looping out of the normal helix depends on the temperature, the identity of the bulge nucleotide and the sequence of the clamping basepairs in the duplex surrounding the bulge [35,36]. A single purine base tends to stack in the helix, while a pyrimidine base can be found at either end of the equilibrium, i.e. stacking or looping out. It should be noted that, to date, the flipped-out base has only been found in the major groove of the helix. Thermodynamic studies indicate that for a single-base bulge, a pyrimidine base is on average 0.5 kcal·mol^{-1} more stable than a purine base [37]. Bulges can cause kinks and bending in the DNA helix [38], and it has been shown that DNA intercalators, such as ethidium, much prefer single bulge regions than normal duplex regions [39,40].

In light of the role played by dynamic structures such as nucleic acid bulges in life processes, we initiated an effort to probe for such structures using the enediyne antibiotic NCS-chrom. Herein, we summarize this laboratory's efforts to study the interaction of bulged nucleic acids with the enediyne antibiotic neocarzinostatin. This review is intended to pinpoint the usefulness of this enediyne antibiotic as a structural probe of a broad portfolio of nucleic acid structures and to provide future models for the study of the interaction of these nucleic acid structures with small molecules.

2. Bulge interaction with thiol-activated neocarzinostatin

A. Mechanism

NCS-chrom is extremely labile under physiological conditions. Activation of NCS-chrom can be initiated by light, heat, pH, radicals, and nucleophiles such as thiol and NaBH$_4$ [1].

The thiol-dependent activation of NCS-chrom, shown in Scheme 1, involves nucleophilic attack by thiol at C12 in *trans* configuration to the naphthoate at C11, and epoxide opening at C5 to generate the 9-membered ring eneyne-cumulene, which cycloaromatizes between C3 and C7 to form a 2,6-indacene biradical [41]. The biradical abstracts hydrogen atoms from hydrogen sources such as solvent and/or DNA to give the postactivated thiol-NCS-chrom adduct tetrahydroindacene [42].

In the case of duplex DNA, the binding of NCS-chrom to DNA is through intercalation of the naphthoate residue of NCS-chrom via the minor groove and placement of the enediyne-containing moiety so that the generated biradical can abstract hydrogens from the minor groove accessible carbons (C1',C4', and C5') of the sugar backbone of the DNA strands [1]. Molecular oxygen adds onto such carbon-centered radicals to form peroxy radicals that give rise to the final oxidative DNA damage products as shown in Scheme 2. Under anaerobic conditions, a DNA-drug adduct may be formed instead.

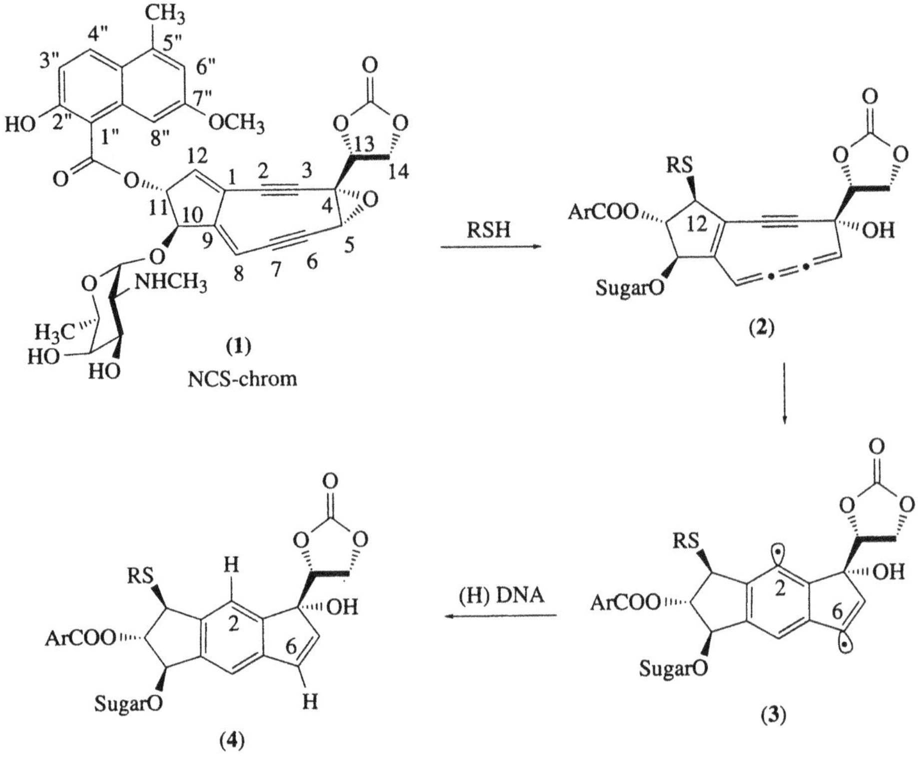

Scheme 1. Proposed mechanism for the thiol activation of NCS-chrom.

B. DNA cleavage specifically at a site opposite to the bulge

In 1988, we developed a "moving lesion" assay to study the interaction of a single-base DNA bulge with intercalation oriented antibiotics [43]. This assay utilizes a series of eight analogous oligodeoxynucleotide duplexes (Fig. 1A). The sequence of the plus strand (D₁) (dodecamer 5′-CGACCCAAATGC-3′) is invariant in eight duplexes. The minus strand of duplex 1 is the dodecamer that is complementary to the plus strand and thus, duplex 1 does not contain a bulge. The minus strands of duplexes 2–8 are each a 13-mer that is complementary to the plus strand with the exception of a single extra cytidine residue. The position of the cytidine bulge has been shifted stepwise in each succeeding member of the series. In this assay, sites of specific binding shift in a stepwise manner in each succeeding duplex of the series, if the bulge site is the specific binding site. It was found that a thiol-activated NCS-chrom consistently causes specific scission on the strand opposite to the unpaired base at the position just 3′ to the single base bulge site, whereas on the bulge-containing strand, the single-strand cleavage at the T site nearest to the bulge is diminished (Fig. 1B). The results showed that, independent

A)

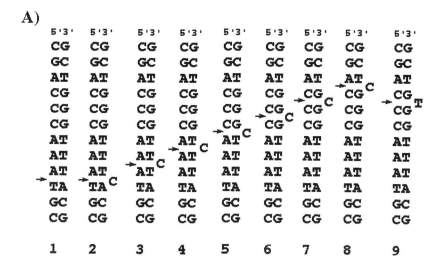

B)

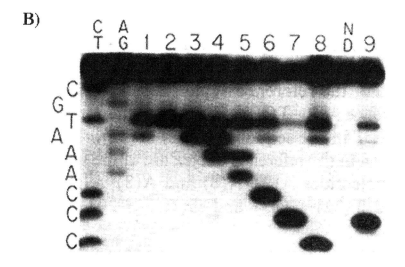

Fig. 1. (A) DNA sequences used in "moving lesion" assay of cleavage of bulged (single-unpaired base) DNA. The arrow denotes the major cleavage site; (B) Sequencing gel of bulged (single-unpaired base) DNA showing "moving lesion" assay. The plus strands of duplexes 1–9 have been labeled at the 5'-end with ^{32}P. C + T and A + G, Maxam and Gilbert sequencing reactions; lanes 1–9, reactions of duplexes 1–9, respectively; ND, (no drug) control reaction [43].

1'-H abstraction
4'-H abstraction
5'-H abstraction
2-deoxyribonolactone
1'-abasic site
3'-phosphate
5'-phosphate
3'-phosphoglycolate
base propenal
5'-phosphate
3'-phosphopyridazine
5'-phosphate

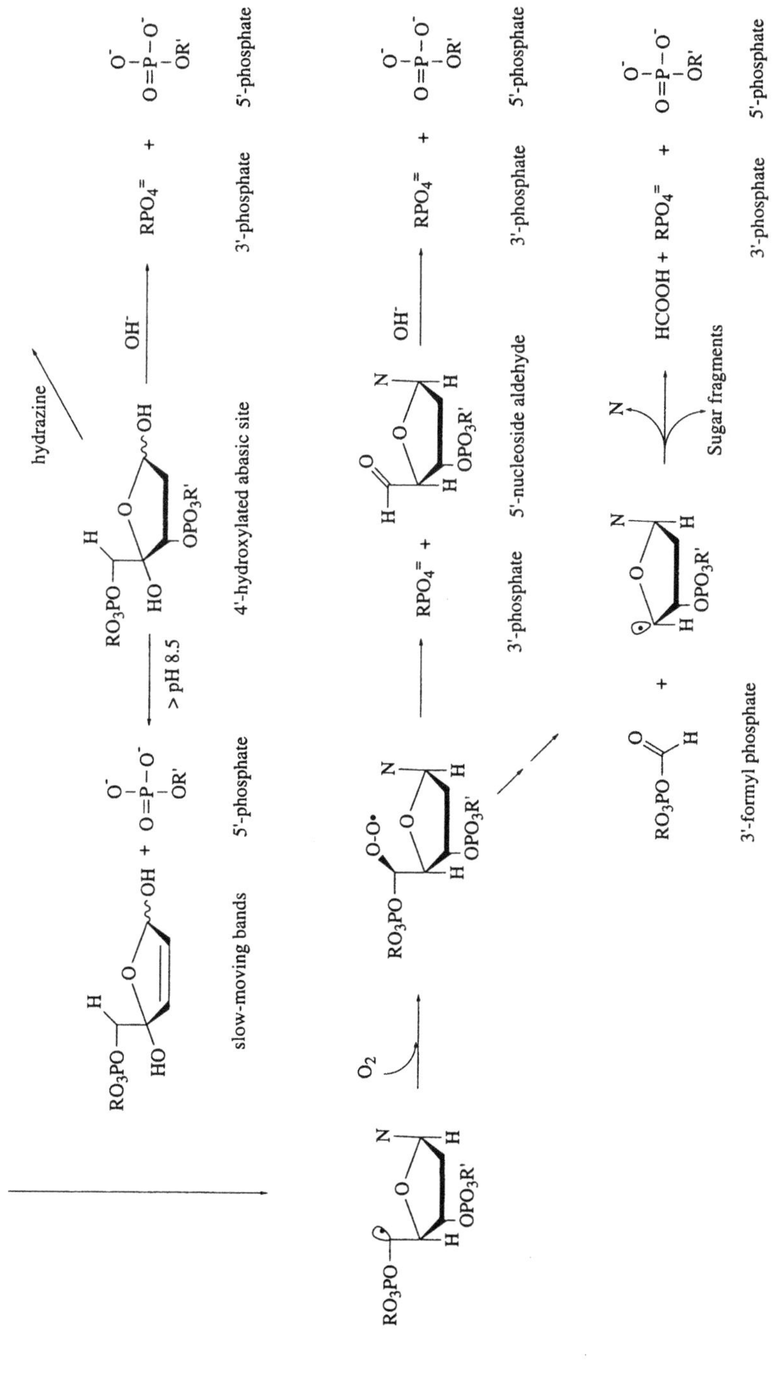

Scheme 2. Proposed mechanisms of oxidative DNA damage.

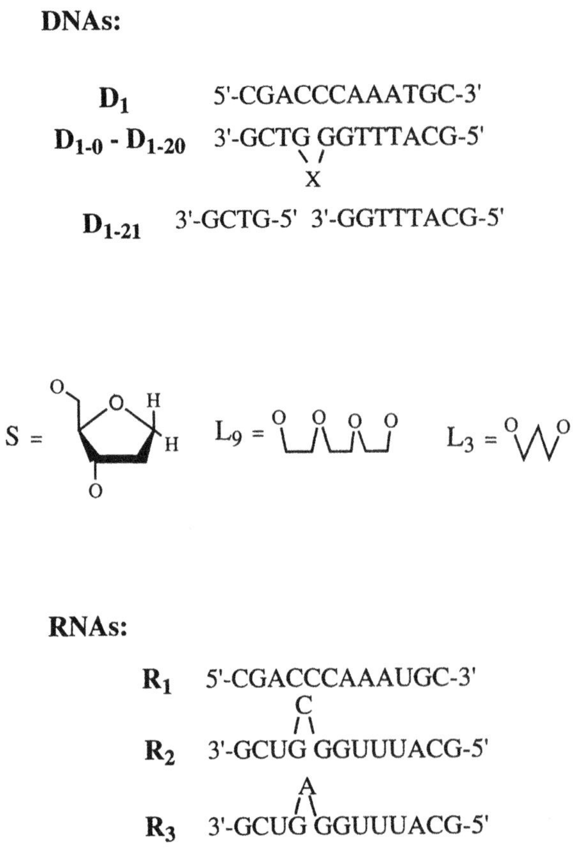

DNAs:

D_1 5'-CGACCCAAATGC-3'

D_{1-0} - D_{1-20} 3'-GCTG GGTTTACG-5'
                  \ /
                   X

D_{1-21} 3'-GCTG-5' 3'-GGTTTACG-5'

Oligonucleotides	
name	bulge bases (X)
D_{1-0}	none
D_{1-1}	C
D_{1-2}	CC
D_{1-3}	CCC
D_{1-4}	CCCC
D_{1-5}	CCCCC
D_{1-6}	T
D_{1-7}	TT
D_{1-8}	TTT
D_{1-9}	TTTT
D_{1-10}	A
D_{1-11}	AA
D_{1-12}	AAA
D_{1-13}	CA
D_{1-14}	AC
D_{1-15}	G
D_{1-16}	S
D_{1-17}	SS
D_{1-18}	L_9
D_{1-19}	L_3
D_{1-20}	L_3L_3
D_{1-21}	strand break

$S = $ $L_9 = $ $L_3 = $

RNAs:

R_1 5'-CGACCCAAAUGC-3'
                  C
                  /\
R_2 3'-GCUG GGUUUACG-5'
                  A
                  /\
R_3 3'-GCUG GGUUUACG-5'

Fig. 2. Sequences of oligonucleotides (names in bold). D_1 and R_1 are non-bulge-containing target strands. D_{1-0} is the perfect complementary strand of D_1. The remaining strands contain extra unpaired base(s) so as to form a bulge in the duplex. D_{1-21} are two strands lacking a connecting phosphodiester linkage so as to form a strand break at the supposed bulge site with D_1 [44].

of sequence, a DNA bulge site is a preferential binding and damaging site by thiol-activated NCS-chrom and that the damage is single-stranded. There was no cleavage in the bulge itself.

In as much as single-base bulge sites are potential drug targets, a more detailed analysis of this reaction was recently carried out [44]. The same plus strand dodecamer (D_1), as discussed above, was chosen as the non-bulge containing target strand (Fig. 2). The single T residue placed in the duplex region is used as an internal control, since it is expected to be a major single-strand cleavage site, based on the study of duplex DNA.

D_{1-0} is a DNA 12-mer that forms a perfect duplex with D_1. All the other strands (D_{1-1} to D_{1-21}) contain an extra base or bases between G(8) and G(9) of the parent strand D_{1-0}. These unpaired bases form a bulge when each of these strands is annealed with D_1, and the bulge will be located directly opposite the site between C(4) and C(5) of D_1. There are five bases between the internal control target (T residue) and the bulge site of D_1, so that interference between the two sites should be minimal. Up to 5 pyrimidine residues were inserted at the bulge site (D_{1-1} to D_{1-9}). For purine bases, bulges of one to three adenines (D_{1-10} to D_{1-12}) or one guanine (D_{1-15}) were studied. Three kinds of chemical linkers resembling the nucleic acid sugar–phosphate backbone were also studied as bulge-mimicking components (D_{1-16} to D_{1-20}). Finally, D_{21} is composed of two DNA strands lacking an interconnecting phosphodiester linkage and thus contains a break between G(8) and G(9) at the original bulge site, when annealed with D_1.

D_1 was 5′ end-labeled with ^{32}P and annealed with various complementary strands (D_{1-0} to D_{1-21}). The duplexes formed were reacted with glutathione-activated NCS-chrom, and after the reaction, each of the mixtures was separated on a 15% denaturing gel. The perfect duplex (D_1 with D_{1-0}) has a strong cleavage band at the T site and a moderate band at

Table 1
Bulge-induced cleavage by thiol-activated NCS-chrom

Oligo	Bulge base	Relative cleavage (%)[a]	Selectivity index (bulge vs. T)[b]
D_{1-1}	C	44	11.7
D_{1-2}	CC	66	5.6
D_{1-3}	CCC	19	1.4
D_{1-4}	CCCC	15	0.5
D_{1-4}	CCCCC	7	0.3
D_{1-6}	T	35	9.4
D_{1-7}	TT	58	4.6
D_{1-8}	TTT	38	1.0
D_{1-9}	TTTT	11	0.3
D_{1-10}	A	100	23.5
D_{1-11}	AA	9	0.73
D_{1-12}	AAA	2	0.2
D_{1-13}	AC	15	1.4
D_{1-14}	CA	56	2.3
D_{1-15}	G	52	5.7
D_{1-16}	S	26	2.8
D_{1-17}	SS	17	1.5
D_{1-18}	L9	10	1.5
D_{1-19}	L3	3	1.2
D_{1-20}	L3L3	5	1.2
D_{1-21}	break	6	0.2
R_2	RNA C	4.7	--
R_3	RNA A	2.1	--

Names of the oligonucleotides complementary to D_1 as described in Figure 2; [a] Relative percentage values of the bulge-induced cleavages in each duplex (intensity percentage of the cleavage band at the bulge opposite the "C" site in each lane) compared to the A bulge, which is normalized to 100%; [b] Ratio of bulge site cleavage versus the internal control (T) site cleavage (intensity ratio of the two bands)[44].

the A site of the normal duplex. Cleavage at these sites exists for all the bulge-containing duplexes, although the intensities vary. The cleavage chemistry was found to involve an initial C5′ hydrogen abstraction [44].

Very strong cleavage occurs across from the bulge structures for almost all the bulge containing duplexes, corresponding to the C (5) position. The bulge formed by a single A base leads to the strongest cleavage on the target strand at the C (5) site. This is not only shown by the highest intensity among the cleavage bands at the same location, but by the highest selectivity (Selectivity index (SI) = 23, Table 1) when compared with the cleavage at the internal control T (and A) site. Table 1 lists the selectivity indices of the bulge-induced cleavages versus the internal T site cleavages. From these results the following conclusions are possible [44]: 1) At the same drug/substrate ratio, for single-base bulges, cleavage is preferred at the bulge site rather than the internal control site in the duplex region (SI > 10, except for G base), and for two or more base bulges the SI drops dramatically (selectivity on bulge-induced cleavage over T site cleavage is 5.6-fold for two C and 0.3 for 5 C); (2) The cleavage specificity at a single-base bulge is A > C > T >> chemical linkers; (3) For the A base, the cleavage induced by a two-base bulge is much weaker than by a single-base, and almost negligible for a three-base bulge; (4) For pyrimidine bases (C and T), two-base bulges lead to cleavage that is about half that of a single-base one in terms of selectivity, and the cleavage is still weaker with three more bases; (5) Cleavage induced by the 5′-CA bulge is almost 4 times that by the 5′-AC bulge, and the SI is also higher for the CA bulge; (6) Because the extra G base at the bulge can cause base slippage in the bulge region, we observe cleavage at all C positions (C(4), C(5), and C(6)) although C(5) is the main cleavage site; (7) Cleavage data from chemical linkers (D_{1-16}–D_{1-20}) show that one abasic site mimic (D_{1-16}) probably retains a significant structural element of the bulge. A break on the complementary strand of D_1 at the supposed bulge location causes moderate cleavage on D_1, but when there is a gap due to deletion of G(8) on the duplex, there is no obvious induced cleavage. This is different from an earlier report showing that dynemicin A induces strand scission at a location adjacent to a one-base gap on the opposite strand [45]. These results imply that a strong drug interaction with the bulge needs more than just the spacious binding pocket. In agreement with earlier results, there was no cleavage in the bulge region itself when the bulge-containing strands were labeled and examined after the reaction.

The observed greater efficiency of cleavage opposite a purine-containing bulge may be due to thermodynamic and/or kinetic factors. Binding studies indicate that kinetic considerations relating to the geometry of the cleavage complex are likely responsible for the difference in cleavage opposite a purine or a pyrimidine containing bulge. It is striking how a simple bulge can induce such strong, specific DNA damage by thiol-activated NCS-chrom. In light of the NMR structures of a complex formed between thiol-postactivated NCS-chrom and a duplex DNA, which shows a minor groove drug binding mode [18,19], and that between base-inactivated NCS-chrom and a two-base bulge DNA, which shows a major groove drug binding mode [46], the binding mode of the drug–DNA interaction induced by a single-base bulge is of particular interest. A preliminary NMR study of a complex formed by thiol-postactivated NCS-chrom and a DNA hairpin containing an A bulge shows similar drug–DNA interaction patterns as observed in the NMR structure of a complex formed between thiol-postactivated NCS-chrom and duplex DNA, suggesting that naphthoate intercalation occurs via the minor groove (unpublished data with X. Gao). The intercalation site appears to be the bulge proper.

C. RNA bulge cleavage

Since bulges are very common in the secondary structures of RNA, it was of interest to see if bulges involving RNA oligonucleotides could be cleaved by thiol-activated NCS-chrom. We synthesized three RNA oligomers, which have similar sequences as D_1, D_{1-1} and D_{1-10} (thymine bases are replaced by uridine bases for RNA, see Figure 2). The cleavage experiments were performed with gluthathione-activated NCS-chrom and annealed bulge-containing DNA–RNA, RNA–DNA hybrids and RNA–RNA duplexes. Interestingly, when DNA is the target strand and RNA is the bulge strand, there is weak cleavage at the usual T and A sites, and the intensities for these bands are comparable with their DNA counterpart. On the other hand, for the bulge-induced cleavages, there is very weak cleavage corresponding to the C(5) position with the RNA C (R_2) bulge, and with the RNA A (R_3) bulge this is even fainter. This is in contrast to DNA bulges where A > C. The intensity for lesions induced by the RNA C bulge is less than 10% of the one by the DNA C bulge. On the other hand, when RNA is the target strand and the bulge strand is either DNA or RNA, no cleavage was observed [44].

It is likely that the dramatic difference in substrate properties between DNA and RNA is due to the inherent structural and conformational differences between these two types of biopolymers. DNA duplexes prefer B-form as their major conformation in aqueous solution, in which the furanose rings are in C-2′-endo puckered conformation. The furanose ring of RNA has stronger preference for C3′ endo puckering [47]. As a result, RNA duplexes and DNA–RNA hybrids normally adopt the A-form conformation. Clearly, the binding of the drug could be much different for these two forms. Further, the extra 2′-OH group on RNA could also play a critical role in the binding and/or positioning of the drug at the bulge site for an effective initial hydrogen abstraction. Taken together, these two factors could account for a much weaker cleavage on the DNA strands when they are the target strands in hybrid duplexes and no cleavage at all on the RNA strands in either hybrid or duplex RNA.

D. Binding

The dissociation constants (K_d) for the complex formed between glutathione-postactivated NCS-chrom and DNA duplexes containing a single binding site are in the range of 2 to 30 μM at 5°C [48]. These measurements were based on the quenching of fluorescence (at 440 nm) of the drug upon DNA binding. Given the fact that bulge-induced cleavage is at least 10 times stronger than that at the internal control site (duplex region), it was anticipated that the bulge site would also be a much more preferred binding site for the drug.

To ensure that there is efficient duplex formation at low DNA concentrations, the two complementary DNA strands, one of which contained a bulge, were connected by a linker to form a hairpin-like structure. Accordingly, we designed several hairpin DNA strands, which in general are D_1 plus its complementary strands connected by chemical linker L_9

Table 2
Dissociation constants and cleavage selectivity indices for hairpin duplexes

Entry	Oligonucleotides	Dissociation constant $(\mu M)^a$	Cleavage selectivity index [b]
1	5'-CGACCCAAATGC-L$_9$-GCATTTGGGTCG	0.150	n/a
2	5'-CGACCCAAATGC-L$_9$-GCATTTGG**A**GTCG	0.043	7.9
3	5'-CGACCCAAATGC-L$_9$-GCATTTGG**AA**GTCG	0.053	0.71
4	5'-CGACCCAAATGC-L$_9$-GCATTTGG**C**GTCG	0.041	2.7
5	5'-CGACCCAAATGC-L$_9$-GCATTTGG**CC**GTCG	0.037	2.4
6	5'-CGACCCAAATGC-L$_9$-GCATTTGG**CCC**GTCG	0.140	n/a
7	5'-CGACCCAAATGC-L$_9$-GCATTTGG**T**GTCG	0.047	n/a
8	5'-CGACCCAAATGC-L$_9$-GCATTTGG**TT**GTCG	0.031	n/a

The bulge-containing strand was connected to its otherwise complementary strand by L$_9$ (L$_9$ = O(CH$_2$CH$_2$O)$_2$CH$_2$CH$_2$O–) to form a hairpin-like structure. The bulge bases are shown in bold. [a] Binding dissociation constants (μM) as measured by fluorescence quenching with glutathione-postactivated NCS-chrom and derived from curve-fitting with Kaleidagraph software. [b] Cleavage selectivity index for bulge-induced cleavage versus that at the internal control site (T10) measured from the reaction of thiol-activated NCS-chrom with various hairpin duplexes [44].

(Table 2). The dissociation constants measured by fluorescence quenching necessarily reflect the availability of multiple competing binding sites on the formed duplex. The binding data at high DNA/drug ratios and low DNA concentrations, however, represent the preferred binding at the strongest binding site. Table 2 shows the K_d values for the tight-binding region, obtained from curve fitting of the data (the fitting curves reflect the single binding site theorem). Interestingly, the K_d values for one- or two-base bulge-containing hairpin duplexes are = 0.050 μM at 3°C; while the binding constant for the perfect hairpin duplex is close to 0.15 μM, with similar results for the bulge containing three cytosine bases. It is significant that the binding of the drug to bulge-containing duplexes having one or two unpaired bases is only about 3 times stronger than that to the perfect duplex. Also, the binding does not distinguish between purine and pyrimidine unpaired bases in the bulge. While the adenine base bulge showed the highest cleavage efficiency, the K_d is little different from that for bulges containing pyrimidines. Similarly, the bulge with two adenine bases has a K_d value close to that of single A base bulge, although there is a large difference between them in terms of cleavage (Table 1). The cleavage of the hairpin duplexes shows a pattern similar to the duplexes formed by two separate strands. A single adenine base bulge in the hairpin gives the strongest cleavage with a selectivity index of 8, and the other three bulges have a selectivity index of 0.7 (AA) and about 2.5 (C and CC) (Table 2).

The differences between the binding and cleavage data may reflect the fact that the cleavage efficiency is largely controlled kinetically instead of thermodynamically. In considering the cleavage data, two important issues associated with efficient cleavage need to be considered. One is how strongly the NCS-chrom preferentially binds to the bulge site (thermodynamic factor), and the other is how efficiently the radical can abstract hydrogen from the deoxyribose (kinetic factor, based on stereochemical considerations).

The actual cleavage data reflect the combination of these two factors. The relatively lower selectivity in binding and high selectivity in cleavage indicate that the kinetic factor must be playing the major role during the reaction and likely reflect the geometry of the complex involved in hydrogen abstraction.

3. Bulge interaction with base-activated neocarzinostatin

A. Mechanism

In the absence of thiol, duplex DNA is not a target for damage by NCS-chrom, and the drug decomposes in a base-catalyzed reaction under physiological conditions (pH >6) to a mixture of inactive forms [11] In 1993 it was observed that an oligodeoxynucleotide containing a two-base bulge can be highly efficiently and specifically cleaved by NCS-chrom at the bulge site in the absence of thiol at pH > 6. Cleavage was restricted to a target nucleotide at the 3′ side of the bulged DNA and was entirely due to 5′-chemistry, with the formation of oligonucleotide fragments with 3′-phosphate and 5′-nucleoside 5′-aldehyde ends [49]. Based on the determination of the structures of the base-postactivated NCS-chrom drug products in the presence and absence of bulged DNA [50–52], a base-catalyzed intramolecular activation mechanism of NCS-chrom was proposed, as shown in Scheme 3.

In this mechanism, the spirolactone cumulene **5** is generated stereospecifically *via* a general base-catalyzed intramolecular Michael addition at C12 by the enolate anion **1b**, which is a resonance form of the naphtholate anion **1a** of NCS-chrom **1**, resulting in the formation of the 2,6-didehydroindacene biradical **6**. This cascade of reactions occurs spontaneously, and in the absence of bulged DNA the biradical is quenched by other hydrogen sources, such as methanol in the solvent, to yield spirolactones **9**. It is of particular interest that the cyclospirolactone **8** is generated virtually only in the presence of substrate bulged DNA and is the only drug product to contain ^{3}H abstracted from the 5′ position of the targeted bulge nucleotide. Thus, it has been proposed that in the presence of bulged DNA, the spirolactone cumulene **5**, which has been implicated as the binding species that searches for the favored DNA binding site, is in equilibrium between bound and free forms, which lead to cyclospirolactone **8** and to spirolactones **9**, respectively, *via* spirolactone biradical **6**.

Mechanistic study of the base-catalyzed activation of NCS-chrom in the absence and presence of bulged DNA shows that nucleobases in the DNA bulge are not required to form an effective bulge pocket but enhance the binding of the wedge-shaped activated drug molecule [52]. Analysis of solvent deuterium isotope effects on NCS-chrom degradation, and DNA cleavage efficiency experiments suggest that the spirolactone biradical **6** is a relatively stable species and that intramolecular quenching of the C2 radical of **6** to form the biologically active cyclospirolactone radical **7a** occurs first (pathway a in Scheme 3), leaving the C6 radical to abstract the hydrogen atom from the DNA deoxyribose and to form the cyclospirolactone **8**. Binding of the activated drug at the bulge site is required, but not sufficient, for efficient **8** formation, whereas cleavage of bulged DNA is not essential. Efficient generation of **8**, but inefficient DNA damage, comes mainly

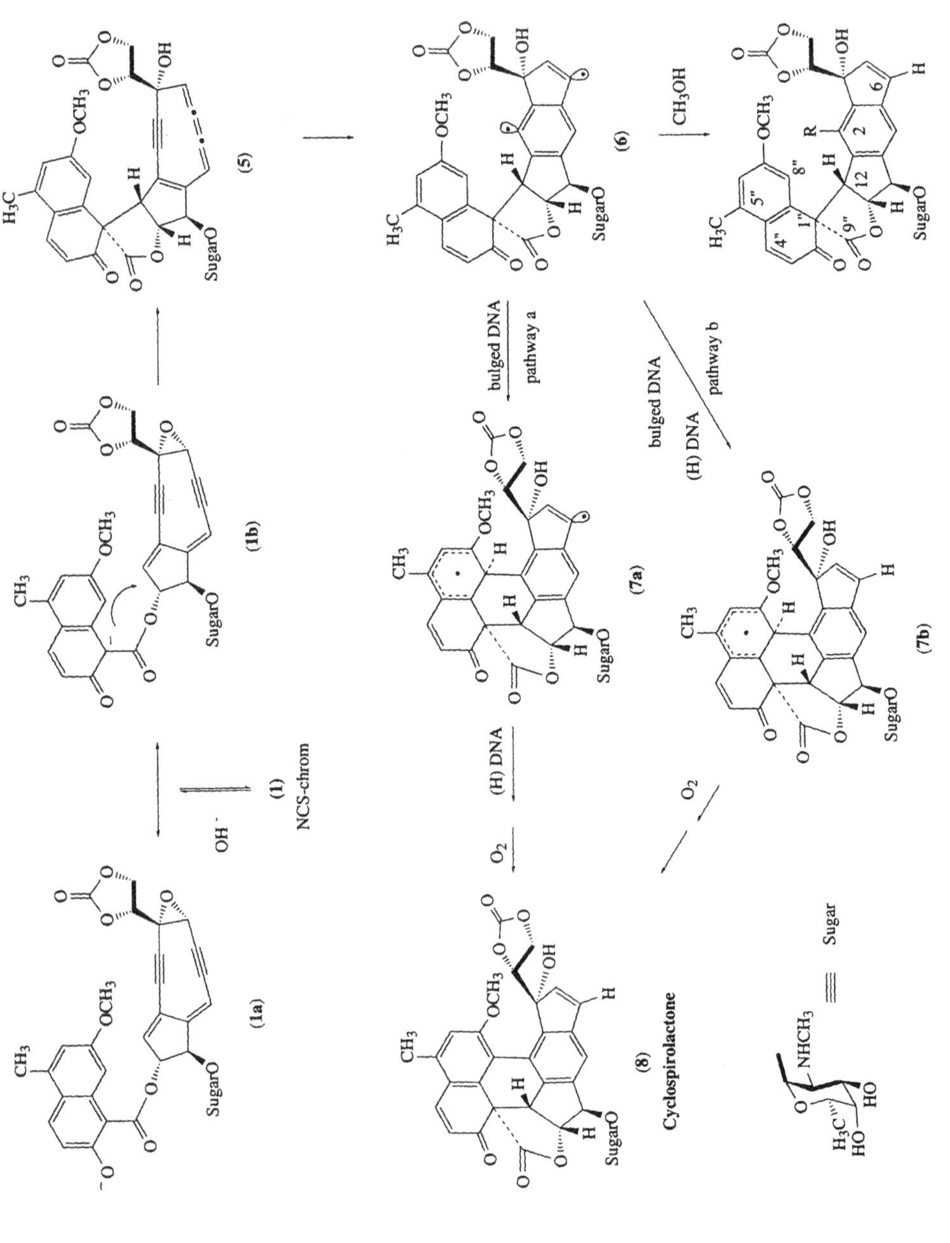

Scheme 3. Proposed mechanistic pathways of base-catalyzed activation of NCS-chrom.

from the likely high off-rate of **7a** binding. The finding that thymidine 5′-carboxylic acid-ended oligonucleotide fragments can be formed in the reaction suggests that the process of DNA cleavage is rather slow and that sequential oxidations of the target 5′-carbon are possible. A study of the effect of solvent (methanol) concentration on NCS-chrom degradation indicates that bulged DNA acts to assist the intramolecular quenching of the radical at C2 by C8′ of the naphthoate moiety by excluding solvent from the binding pocket, thus preventing the formation of spirolactones **9**, and by blocking radical polymerisation [52]. Since in the absence or near absence of methanol, **8** formation does not reach even the 10% that formed in the presence of bulged DNA, it is possible that the DNA bulge also induces a conformational change in the drug to promote the intramolecular reaction. CD studies on the complexation of the base-postactivated NCS-chrom **9a** and a bulged DNA indicate that not only does DNA change conformation due to complex formation, but so does the drug (unpublished data with C. Yang). Base-postactivated NCS-chrom **9a** also induces and enhances the formation of the bulge structure (unpublished data with X. Gao).

B. DNA cleavage specifically at the bulge site

Treatment of single-stranded DNA that corresponds to the 3′-terminus of yeast transfer RNA (tRNA[Phe]) with NCS-chrom in the absence of thiol resulted in cleavage at a single site, T22 (Fig. 3) [49]. Point mutations, deletions and insertions in the DNA analog and its complement of the 3′-terminus of yeast tRNA[Phe] show that for a single-stranded DNA to be cleaved by NCS-chrom in the absence of thiol, the DNA must generate a hairpin structure with an apical loop and a bulge containing at least 2 unpaired bases [49,53]. The cleavage target is the bulged-out nucleotide at the 3′ side of the bulge. The size of the loop is not critical so long as it contains at least three nucleotides; the bulge requires a minimum of two nucleotides but must have fewer than five (Fig. 4). With a notable exception involving base-pair changes immediately 3′ to the bulge, base changes in the bulge and base-pair changes immediately 5′ to the bulge retain substrate activity for NCS-chrom. Conversion of the T23·A12 base pair 3′ to the bulge to G·C (or to A23·T12) eliminates the substrate property; further, the G·C change results in the absence of **9a** binding. Maintenance of the bulged structure requires stable duplex regions on each side of the bulge. A similar bulged structure, lacking a loop, formed by the annealing of a linear 8-mer and a 6-mer is also an excellent target for cleavage by base-activated NCS-chrom. The reason for the diminished cleavage of a single-base bulge or a bulge with more than three bases is quite evident from the NMR solution structure of a complex of **9a** and a two-base bulged DNA (*vide infra*) [46]. In the absence of dioxygen, strand cleavage is blocked and quantitatively replaced by a drug-DNA adduct covalently linked at the expected cleavage site [49,53,54–56].

To make use of the DNA bulge-specific, base-catalyzed NCS-chrom reaction as a probe of the secondary and tertiary structures of naturally occurring bulged structures and to identify new substrate patterns of the reaction, high molecular weight single-stranded phage DNA and long fragments derived therefrom were tested as potential substrates. It was found that single-stranded circular φχ174 and M13mp18 phage DNAs are substrates for the base-activated NCS-chrom, presumably due to bulge formation at certain limited

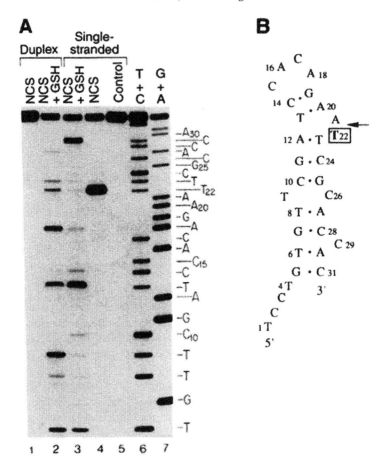

Fig. 3. NCS-chrom-induced bulged (2 unpaired bases) DNA cleavage. (A) Strand cleavage of 5′-³²P-labeled 31-I DNA in the absence or presence of 3 mM glutathione (GSH). Reactions contained double-strand DNA without and with GSH (lanes 1 and 2); single-strand DNA with GSH, without GSH, and without drug or GSH (lanes 3 to 5); and T + C and G + A sequence markers (lanes 6 and 7). (B) Sequence and proposed conformation of 31-I DNA. Nucleotides correspond to positions 47 to 76 in yeast tRNA^Phe, which includes the T stem loop. The arrow indicates the site of cleavage at T22 [49].

sites [57]. These sites were shown to be either pure bulges (Fig. 5B) or bulges (really bubbles) also containing an unpaired base on the opposite strand in the bulge region (Fig. 5A) [57].

In the DNA analog of TAR (transactivation responsive region) RNA of HIV virus, as well as in a DNA duplex made of two linear oligomers that can form a similar bulge, base-activated NCS-chrom causes strand damage at the T residues in the bulge and at the bases flanking the bulge [58]. Cleavage at T25 in the bulge also involves, in addition to 5′-chemistry, 4′ attack. Experiments using DNA substrates having deuterium selectively at the 4′- or 5′-position of T25 show kinetic shuttling between the two positions.

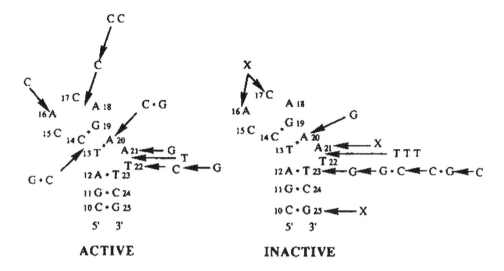

Fig. 4. Summary of the point mutations, insertions, and deletions determining whether oligonucleotide 10–25 of the 31-I in Figure 3 is an active or inactive substrate of base-activated NCS-chrom. Placement position of the arrows indicates whether base(s) is a substitution or insertion. Deletion is indicated by x [53].

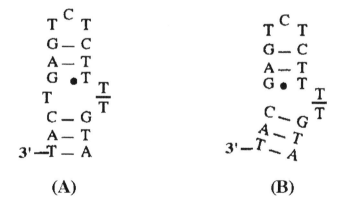

Fig. 5. Oligonucleotide sequences derived from long, single-stranded DNA identified as active or inactive substrates for base-activated NCS-chrom. Underlining denotes the cleavage site [57].

Sequence changes in TAR DNA show that the specificity and extent of cleavage is sequence-dependent. (Fig. 6)

The base-catalyzed NCS-chrom cleavage reaction is 50-fold slower than the thiol-activated reaction. This probably is reflective of the activation step, which for glutathione in the presence of DNA is very fast [59]. Nevertheless, the thiol-independent, site-specific cleavage in the bulged DNA is highly efficient and selective. The cleavage efficiency of

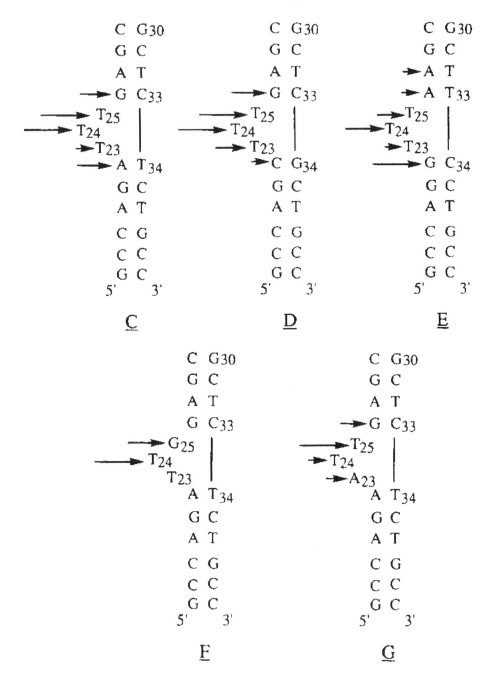

Fig. 6. Effect of sequence changes in TAR DNA on NCS-chrom-induced strand cleavage in the absence of thiol. Arrows point to the site of cleavage. The length of arrows indicates approximately the relative intensity of the gel bands. The radioactivity present in C for all the cleavage bands combined was 5% of the total [58].

the drug is more than 50% [52]. Unlike in the thiol-dependent reaction with duplex DNA, where high levels of drug cause breaks at less favored sites, no other cleavage sites but at the bulge are found in the two-base bulged DNA in the base-catalyzed reaction. This lesion also differs substantially from the cleavage pattern at the site of a single base-bulge in duplex DNA in the presence of thiol as discussed earlier. In the latter reaction a break is generated on the strand opposite the bulged base at the residue just 3' to the bulge; the bulged residue is not a cleavage target. The difference comes from the different DNA binding modes of the drug; in the base-catalyzed reaction intercalation of the double-decker intercalator **9a** occurs via the major groove [46]; the other is due to single-decker intercalation by the naphthoate residue of the thiol-postactivated NCS-chrom via the minor groove [19].

C. Specific cleavage also occurs at the bulge site in RNA

As the genomic RNAs of many RNA viruses have hairpin stem-loop conformations with bulged bases serving as binding sites for viral regulatory proteins, a functional bulge binder would be highly desirable for the study of the regulation of viral RNA synthesis by TAR and RRE (rev responsive element) of HIV virus in AIDS. Given the resemblance in the structural features of HIV-1 TAR RNA to that of the competent DNA substrates, in having a folded structure with an apical loop and a bulge with flanking double-stranded regions stabilized by Watson-Crick base-paired regions, TAR RNA appears to be a good candidate as a potential substrate of base-activated NCS-chrom. Accordingly, a 29-mer (residues 17–45) oligoribonucleotide corresponding to the Tat (transactivator protein) binding portion [60] of the 59-residue TAR RNA was studied. Base-activated NCS-chrom generates a distinct but weak band at the bulge site on a sequencing gel [58]. This band moves slightly faster than that of the band generated by *Bacillus cereus* ribonuclease at U24; in comparison with the alkali-induced cleavage, it appears to have a 3'-phosphate end.

In an effort to determine why bulged RNA substrates are not as good substrates for base-activated NCS-chrom as bulged DNA substrates, systematic substitutions of ribonucleotide residues in a good substrate DNA bulge (CCGATGCG·CGCAG<u>T</u>TCGG) (cleaved residue is underlined) were carried out (Table 3) [55]. It was found that ribonucleotide substitution at the bulge target site, as well as at other regions involving duplex formation, had a small effect on the cleavage reaction, unless either of the two strands was entirely of the ribo form. Substitution of the target T with ribo U decreases cleavage by only 25%; the duplex having both bulge residues in ribo form is cleaved somewhat less efficiently, but still significantly. Of the two A·T base pairs on either side of the bulge, changing the A·T base pair on the 5' side of the target bulge nucleotide (underlined T residue) to ribo A·U resulted in an 87% decrease in cleavage. Conversion of the A alone to the ribo form caused a 68% loss in cleavage. On the other hand, conversion of the A·T on the 3' side of the target bulge nucleotide T to the ribo forms alters cleavage to a much lesser extent. It is interesting to note that the hybrid having all but the central core of six residues inclusive of the two-base bulge in the ribo forms still yields 44% cleavage at the bulge. These results can be understood from the NMR solution structure [46] of the complex formed between the base-postactivated NCS-chrom, an isostructure of the base-activated NCS-chrom biradical, and a bulged DNA, since a projected 2'-hydroxyl group

Table 3
Effect of ribonucleotide substitution on cleavage

Substrate	Duplex	Cleavage (%)
1	5′ -CCGATGCG 3′ -GGCTACGC T̲G	100
2	5′ -CCGATGCG C′ -GGCTACGC u̲G	75
3	5′ -CCGATGCG 3′ -GGCTACGC u̲g	60
4	5′ -CCGaTGCG 3′ -GGCuACGC T̲G	66
5	5′ -CCGAuGCG 3′ -GGCTaCGC T̲G	13
6	5′ -CCGATGCG 3′ -GGCTaCGC T̲G	32
7	5′ -ccgaugcg 3′ -GGCTACGC T̲G	4
8	5′ -CCGATGCG 3′ -ggcuacgc u̲g	2

The data are from the gel electrophoresis. 100% represents 77% cleavage in the duplex composed of deoxyribonucleotides only. Lower case letters denote ribo residues. Cleavage occurs at the underlined and bold **T̲** or **u̲** residue [55].

of the ribo A at the 5′ side of the bulge would be expected to clash sterically with the 7′-methoxyl of the drug. These results suggest that deletion of the 7-methoxy moiety would lead to better binding for the RNA bulge.

These studies show that a well-defined bulged RNA structure can be specifically targeted by enediyne antibiotics of the NCS-chrom type and provide a working model for designing a new generation of enediyne-based molecules that specifically and efficiently target the bulge region of an RNA.

D. Binding and NMR structure

In an effort to identify the active species responsible for hydrogen atom abstraction from the bulge and to obtain a three-dimensional structure of its complex with bulged DNA, we examined the binding of stable drug metabolites, generated in the course of the cleavage reaction, with oligodeoxynucleotides containing the bulged structure. By use of fluorescence quenching, it has been found that one drug product, **9a** (excitation at 390 nm, emission at 500 nm), which is formed in the absence of bulged DNA and most closely resembles the biradical intermediate **6** in the cleavage reaction, specifically binds bulged DNA with a K_d in the low micromolar range and competitively inhibits the cleavage reaction [61]. By contrast, the DNA bulge-specific drug metabolites **8** (excitation at 400 nm and emission at 530 nm) and **9b** (excitation at 390 nm and emission at 500 nm) showed no fluorescence quenching by the bulge-containing oligonucleotide. Using a series of oligonucleotides containing different bulged structures, it has been found that: a) the binding profile resembles the cleavage profile; strong binding corresponds to strong cleavage; and b) the cleavage reaction is more sensitive than the binding of **9a** to bulged DNA, as measured by fluorescence quenching. This is mainly due to the stability of the presumed bulge structures, as some short two-strand constructs of bulged structures, with weakly observed fluorescence quenching, still give strong bulge cleavage.

NMR studies on the complex of the spirolactone **9a** and a two-base bulged DNA (5′-CCCGATGC-$(OCH_2CH_2)_3$-O-GCA<u>AT</u>TCGGG-3′, underlined A and T denote the bulge bases after the oligonucleotide is folded into a hairpin structure) reveal that the spirolactone functions as a double-decker intercalator (connected by the spirolactone ring) with the tetrahydroindacene moiety stacking with the base-pair above and the naphthoate moiety stacking with the base-pair below the bulge [46]. The wedge-shaped drug fits tightly into the triangular prism pocket formed by the two looped-out bases and the

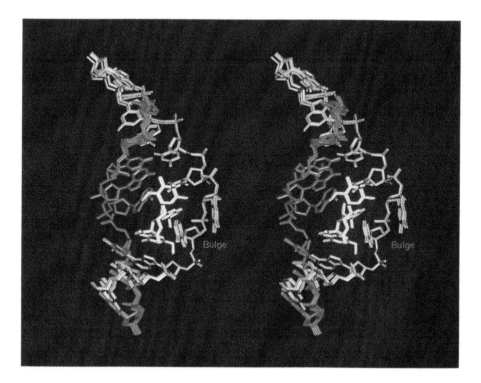

Fig. 7. Overlay plot of the six NMR-derived structures of the complex between base-postactivated NCS-chrom and a bulge-containing oligonucleotide in stereoview (direct eye) through the major groove. The bulge residues A12 and T13 are on the right side [46].

neighboring base pairs via the major groove. The drug carbohydrate sits at the center of the major groove, providing the needed charge neutralization to enhance the binding and limiting further drug penetration. The two drug rings mimic the geometry of helical DNA bases with a twist angle of 35 degrees and a rise of 3 Å at the narrower end, and mediate the helical transition between the two half helices on either side of the bulge, complementing the bent DNA structure. The cyclocarbonate residue in tetrahydroindacene actually makes the stacking base-pair T·A buckle, which may provide the molecular basis for the observed weakened binding and cleavage in the case of reversing the T·A base-pair at the 3' side of the bulge. The putative abstracting drug radical is 2.2 Å from the pro-S H5' of the target bulge nucleotide. It should be noted that base-postactivated NCS-chrom **9a** actually induces and enhances the formation of the bulge structure (unpublished data with X. Gao). Figure 7 shows the overlay plot of the six structures of the complex in direct eye stereoview through the major groove. This structure clarifies the mechanism of DNA bulge recognition and cleavage by NCS-chrom and provides insight into the design of bulge-specific DNA/RNA binding molecules.

4. Designing synthetic bulge-specific binders

A. Synthesis

Much effort has been devoted to the design of specific binders targeting the bulge motif, especially of RNA, due to its biological importance [62,63]. The specific strong binding of the base-postactivated NCS-chrom **9a** to bulge DNA [61] and the determination of the NMR solution structure [46] of such a binding complex enable us to understand and develop more stable and functional bulge binders. Chemical synthesis of accessible analogs of **9a** was pursued in an effort to identify the minimum structural requirements for an effective DNA bulge binder with the ability to undergo double-decker intercalation (DDI). As discussed before, a key structural feature of **9a**, which is presumed to play a vital role in DNA bulge recognition, is the right-handed helical twist imposed by the spirocyclic junction, mimicking the right-handed helix in DNA. Scrutinizing this parameter using semiempirical calculations with a range of potential structural isosteres suggested that a simple mimic of this system could be afforded by spiro-alcohols **13** and **19**. Functional group transformations applied to **13** and **19** would offer access to a wide variety of analogs and also provide a platform to modulate stereochemical bias. The synthetic routes are depicted in Schemes 4 and 5 [64].

B. Binding

These designed synthetic double-decker intercalators were found to be bulge- specific binders. They interact with bulged DNA in the low micromolar range. Table 4 lists the dissociation constants (K_d) of these compounds bound to a model DNA bulge. By attaching a sugar or an aminosugar moiety at one of the readily available hydroxyl groups in these compounds, water-solubility of the synthetic double-decker intercalator has been improved dramatically which facilitates the binding. Of significance is the finding that

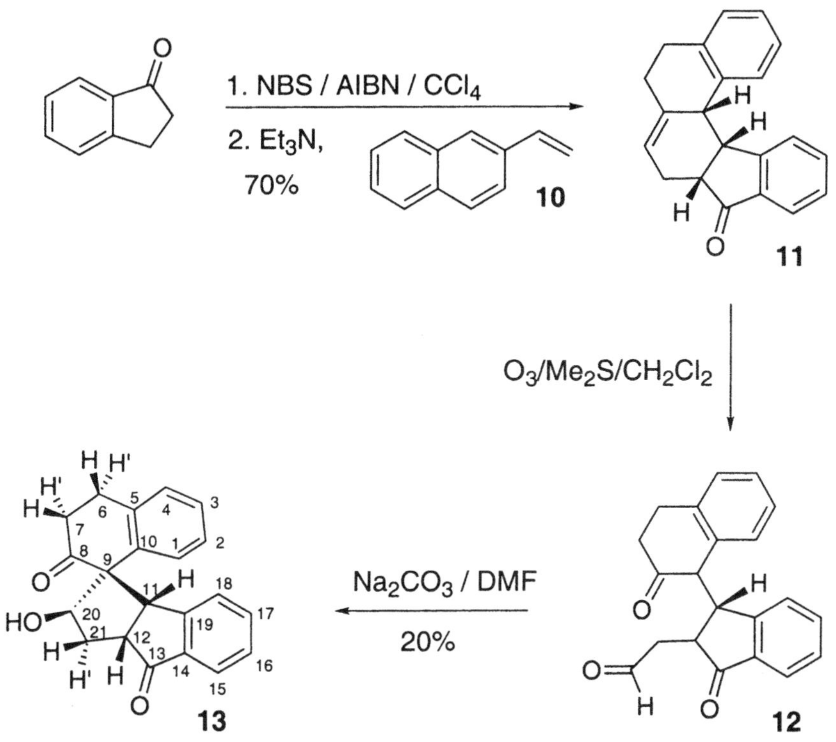

Scheme 4. Synthesis of pentacyclic DDI-A.

reversal of the T·A base-pair at the 3′ side of the bulge, which creates a poor binding site for spirolactone **9a** from NCS-chrom, leads to a binding site for the DDI compounds that is as good as the original (unpublished data with G. Jones). Expansion of the core DDI, such as combining with other reactive functional groups (alkylating group, free radical generating moiety, hydrolyzing group such as metal ion etc.), or/and adjusting the aromatic arena to accommodate better base-pair selectivity should lead to improved tools for the study of the biology of bulged nucleic acid structures and their role in disease states. We are currently pursuing these in collaboration with the laboratory of Professor Graham Jones at Northeastern University.

5. Adduct formation

Since the DNA-bound postactivated enediyne core is an unsaturated system, the newly generated proximal sugar radical has the ability to add back onto the unsaturated system to form a covalent adduct or even an interstrand cross-linked product. The actual process is one of kinetic differentiation, depending on the relative position of the sugar radical

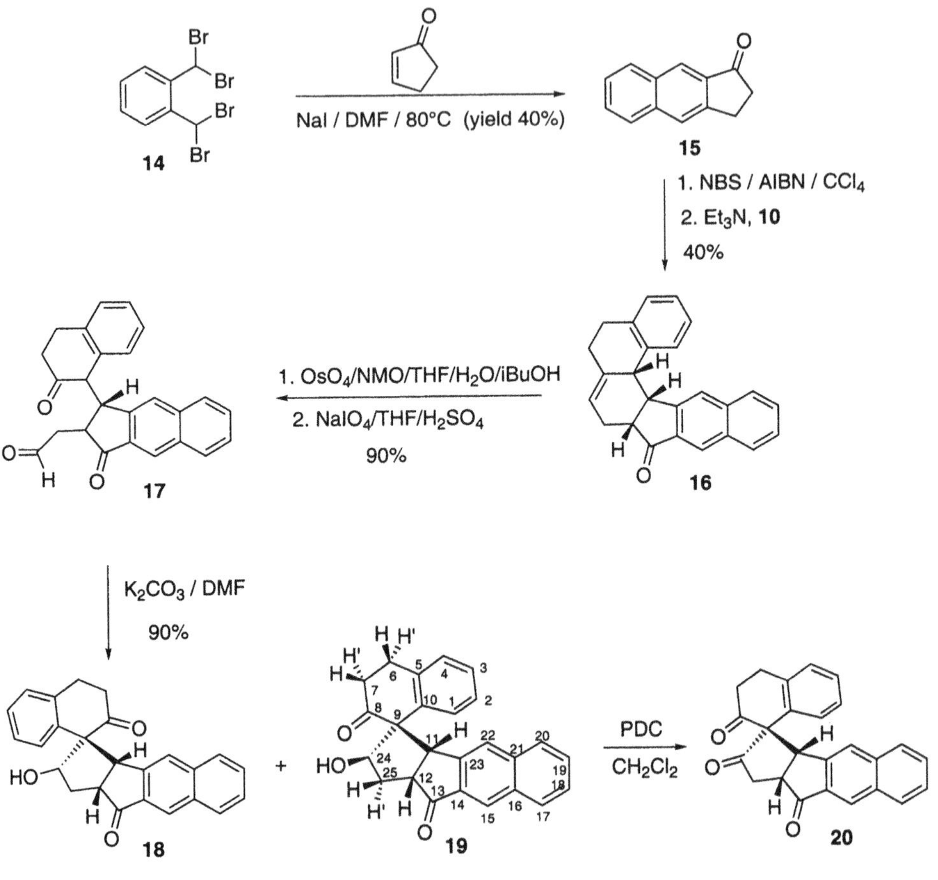

Scheme 5. Synthesis of DDI-B.

to the unsaturated system and the ease of translocation of the sugar radical by an intramolecular or intermolecular process, such as quenching by dioxygen or other hydrogen sources, such as solvent or thiol. Adduct formation is competitive with dioxygen quenching. The formation of novel drug-DNA adducts involving the DNA deoxyribose moiety has been observed for both duplex and bulged DNA, especially under anaerobic conditions, in both thiol-activated and base-activated NCS-chrom reactions [49,53–56, 58,65–68]. A drug-deoxyribose adduct, involving C5' of deoxyribose of DNA, has been isolated. As found for the thiol-activated NCS-chrom reaction with single-base bulges, the base-catalyzed cleavage reaction is also relatively independent of the surrounding DNA sequence, except, as noted for the base pair on the 3' side of the bulge. Resembling the DNA damage behavior in bulged DNA, the site of adduct formation is relatively independent of the DNA sequence involved in thiol-activated NCS-chrom reactions [44].

Table 4
Affinity of NCS-chrom cascade mimics for bulged DNA sequences [64]

Compd	Excitation (nm)	Emission (nm)	Bulge[b] K_d (μm)	duplex[c] K_d (μm)
9a	390	500	2.2	307
±**-13**	nd[a]	nd[a]	≈50[a]	
±**-19**	310	450	25	90
*ent*₁**-19**	310	450	35	89
*ent*₂**-19**	310	450	17	90
±**-18**	310	450	>500	

[a] Competition assay against **9a**.
[b] Bulge DNA: 5'-GTCCGATGCGTG-3'
                        3'-CAGGCTACGCAC-5'
                                TG
[c] Duplex DNA: 5'GTCCGATGCGTG-3'
                        3'CAGGCTACGCAC-5'

The observed partial strand breakage by piperidine treatment of the isolated adduct band isolated from a sequencing gel, from both thiol-activated and base-activated NCS-chrom reactions with bulged DNA under anaerobic conditions [44,54], indicates that the adduct formed under these reaction conditions is actually a mixture of at least two types. Only about 10–20% of the adduct is alkali-labile. The breakdown gives either a fragment with a 5' phosphate or a fragment with a 3' phosphate. It is plausible that the two species of drug-DNA adduct come from the same biradical drug intermediate, through the addition of the newly generated 5' carbon deoxyribose radical to the double bond involving the newly quenched C6 radical of the drug. This would lead to either an adduct containing a single bond between C5 and C6 (single-bond adduct) or to one containing a double bond between C5 and C6 (double-bond adduct). The latter could also come from radical collision between the newly formed C5' radical and the second incoming biradical drug intermediate. The double-bond adduct would be expected to be base-labile and lead to the breakage of the nucleotide chain upon reacting with hot piperidine by two successive β-scissions, forming both the 3' and 5' phosphate-ended fragments (Scheme 6). Structure determination of these drug adducts should shed further light on the precise mechanism of this type of DNA damage. The ability to generate a site-specific deoxyribose-drug adduct at a bulge in DNA, regardless of the sequence involved, should provide a useful tool for the study of the cytotoxic effects caused by these types of adducts. The significant increase of drug-adduct formation under anaerobic conditions may have important implications for its cytotoxicity in the central regions of large tumors, where oxygen tension is low.

6. Concluding remarks

NCS-chrom is unique among the enediyne antibiotics in its ability, upon activation by either thiol or intramolecularly in a general base-catalyzed reaction, to generate two

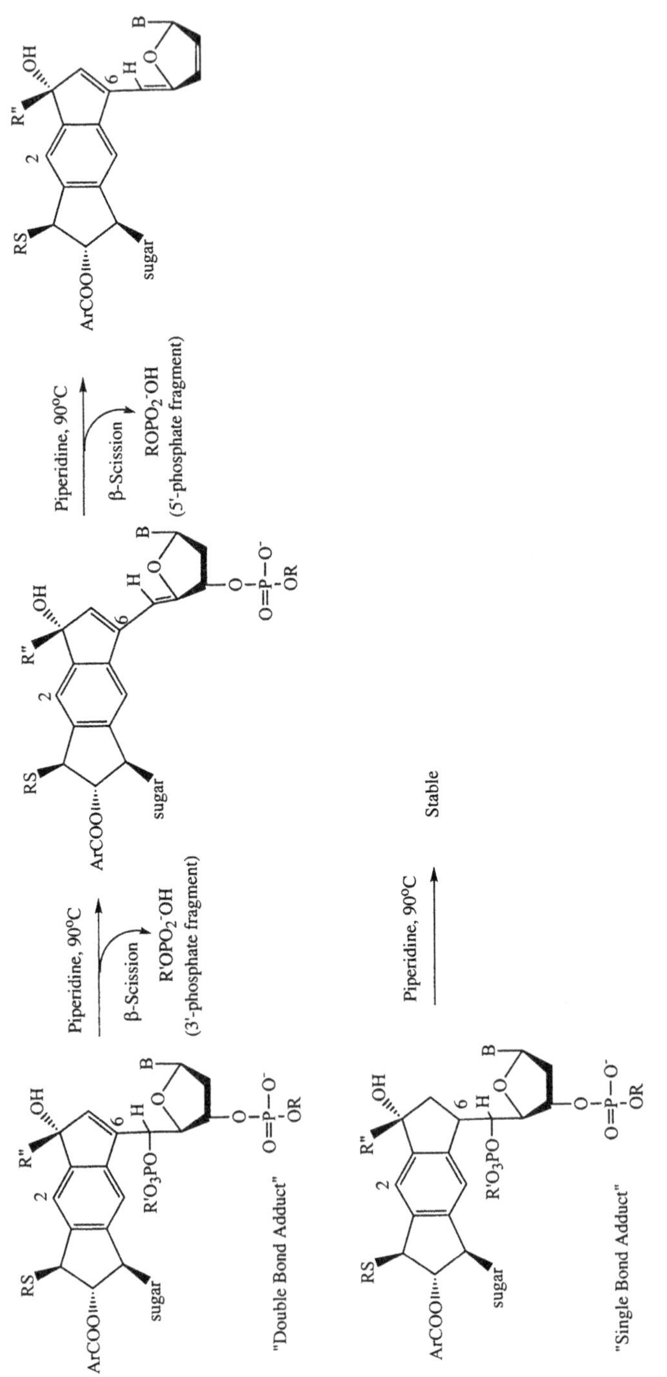

Scheme 6. Possible DNA-NCS-chrom adducts and their alkali degradation.

different DNA cleaving species with different target structures. Due to the presence of a naphthoate moiety, thiol-activated NCS-chrom is converted into an intercalating biradical species selective for single-base bulges in duplex DNA independent of sequence consideration (which otherwise prevails in perfect duplex DNA). Intramolecular activation catalyzed by base, on the other hand, generates a rigid, wedge-shaped (double-decker) molecule that, highly selectively and specifically, interacts with DNA bulges containing two or three unpaired bases. Further, the site of cleavage in the two cases is different, being on the non-bulge strand in one case (thiol) and at the bulge in the other (intramolecular).

Since thiol activation is about 50-fold faster than intramolecular activation and cells are rich in glutathione, it is highly unlikely that significant amounts of the double-decker biradical species would be generated under physiological conditions. Therefore, in order to enable such a species to interact with bulged nucleic acids in the cell, the preformed double-decker molecule would have to be supplied directly to the cell. Undoubtedly, this could be best accomplished by endowing the presynthesized molecule with the ability to interact covalently at the bulge target site by attaching either an enediyne-radical generating moiety or an alkylating group to DDI.

Finally, it is important to point out, as noted earlier, that spirocyclic compounds have the ability to induce the formation of bulged structures in nucleic acids, even when they may not be the preferred structure in single-stranded regions in the absence of the drug. This ability to "induce fit" is due to the rigid drug structure, whose rings mimic helical bases linked together with the proper amount of right-handed twist. It will likely be this property that will determine the ability of such agents to interact with transient slippage-induced bulged structures during the replication of unstable expansions of nucleotide repeats.

Acknowledgement

We wish to thank former and current members of the Goldberg laboratory for their experimental contributions to the field of enediyne antibiotic research and for comments and criticisms of this chapter. This work was supported by U.S. Public Health Service grant GM 53793 from the NIH.

References

[1] Z. Xi, I.H. Goldberg. In: D.H.R. Barton, K. Nakanishi, eds, *Comprehensive Natural Products Chemistry*, vol. 7 (pp. 553–592). Elsevier Science, Oxford, 1999.

[2] In: D.B. Borders, T.W. Doyle, eds, *Enediyne Antibiotics as Antitumor Agents*, Dekker, New York, 1995.

[3] In: B.E. Meunier, ed., *DNA and RNA Cleavers and Chemotherapy of Cancer and Viral Diseases*, Kluwer Academic, Boston, 1996.

[4] N. Ishida, K. Miyazaki, K. Kumagai, M. Rikimaru, *J. Antibiot.* 18 (1965) 68–76.

[5] J. Meienhofer, H. Maeda, C.B. Glaser, J. Czombos, K. Kuromizu, *Science* 178 (1972) 875–876.

[6] Y. Ono, Y. Watanabe, N. Ishida, *Biochim. Biophys. Acta* 119 (1966) 46–58.

[7] M.A. Napier, B. Holmquist, D.J. Strydom, I.H. Goldberg, *Biochem. Biophys. Res. Commun.* 89 (1979) 635–642.

[8] L.F. Povirk, I.H. Goldberg, *Biochemistry* 19 (1980) 4773–4780.
[9] L.S. Kappen, I.H. Goldberg, *Biochemistry* 19 (1980) 4786–4790.
[10] L.S. Kappen, M.A. Napier, I.H. Goldberg, *Proc. Natl. Acad. Sci. USA* 77 (1980) 1970–1974.
[11] M.A. Napier, B. Holmquist, D.J. Strydom, I.H. Goldberg, *Biochemistry* 20 (1981) 5602–5608.
[12] L.S. Kappen, I.H. Goldberg, *Nucleic Acids Res.* 5 (1978) 2959–2967.
[13] I.H. Goldberg, *Acc. Chem. Res.* 24 (1991) 191–196.
[14] P.C. Dedon, Z.W. Jiang, I.H. Goldberg, *Biochemistry* 31 (1992) 1917–1927.
[15] S.M. Meschwitz, I.H. Goldberg, *Proc. Natl. Acad. Sci. USA* 88 (1991) 3047–3051.
[16] S.M. Meschwitz, R.G. Schultz, G.W. Ashley, I.H. Goldberg, *Biochemistry* 31 (1992) 9117–9121.
[17] A. Galat, I.H. Goldberg, *Nucleic Acids Res.* 18 (1990) 2093–2099.
[18] X. Gao, A. Stassinopoulos, J. Gu, I.H. Goldberg, *Bioorg. Med. Chem.* 3 (1995) 795–809.
[19] X. Gao, A. Stassinopoulos, J.S. Rice, I.H. Goldberg, *Biochemistry* 34 (1995) 40–49.
[20] L.S. Kappen, I.H. Goldberg, B.L. Frank, L. Worth Jr, D.F. Christner, J.W. Kozarich, J. Stubbe, *Biochemistry* 30 (1991) 2034–2042.
[21] L.S. Kappen, I.H. Goldberg, *Biochemistry* 31 (1992) 9081–9089.
[22] L.S. Kappen, I.H. Goldberg, *Proc. Natl. Acad. Sci., USA* 89 (1992) 6706–6710.
[23] X. Zeng, Z. Xi, L.S. Kappen, W. Tan, I.H. Goldberg, *Biochemistry* 34 (1995) 12435–12444.
[24] D.H. Turner, *Curr. Opin. Struct. Biol.* 2 (1992) 334–337.
[25] D.M.J. Lilley, *Proc. Natl. Acad. Sci. USA* 92 (1995) 7140–7142.
[26] M. Chastain, I. Tinoco Jr, In: *Progress in Nucleic Acid Research and Molecular Biology*, vol. 41 (pp. 131–177). Academic Press, 1991.
[27] L.S. Ripley, *Proc. Natl. Acad. Sci. USA* 79 (1982) 4128–4132.
[28] G. Streisinger, Y. Okada, J. Emrich, J. Newton, A. Tsugita, I. Terzaghi, M. Inouye, *Cold Spring Harbor Symp. Quant. Biol.* 31 (1966) 77–84.
[29] T.A. Kunkel, *Nature* 35 (1993) 207–209.
[30] V.A. Malkov, I. Biswas, R.D. Camerini-Otero, P. Hsieh, *J. Biol. Chem.* 272 (1997) 23811–23817.
[31] Y.H. Wang, C.D. Bortner, J. Griffith, *J. Biol. Chem.* 268 (1993) 17571–17577.
[32] C.T. Ashley Jr, S.T. Warren, *Annu. Rev. Genetics* 29 (1995) 703–728.
[33] M.W. Kalnik, D.G. Norman, B.F. Li, P.F. Swann, D.J. Patel, *J. Biol. Chem.* 265 (1990) 636–647.
[34] F. Aboul-ela, A.I. Murchie, S.W. Homans, D.M. Lilley, *J. Mol. Biol.* 229 (1993) 173–188.
[35] M.W. Kalnik, D.G. Norman, M.G. Zagorski, P.F. Swann, D.J. Patel, *Biochemistry* 28 (1989) 294–303.
[36] U. Dornberger, A. Hillisch, F.A. Gollmick, H. Fritzsche, S. Diekmann, *Biochemistry* 38 (1999) 12860–12868.
[37] D.A. Leblanc, K.M. Morden, *Biochemistry* 30 (1991) 4042–4047.
[38] J.A. Rice, D.M. Crothers, *Biochemistry* 28 (1989) 4512–4516.
[39] J.W. Nelson, I. Tinoco Jr, *Biochemistry* 24 (1985) 6416–6421.
[40] S.A. White, D.E. Draper, *Biochemistry* 28 (1989) 1892–1897.
[41] A.G. Myers, *Tetrahedron Lett.* 28 (1987) 4493–4496.
[42] A.G. Myers, P.J. Proteau, T.M. Handel, *J. Am. Chem. Soc.* 110 (1988) 7212–7214.
[43] L.D. Williams, I.H. Goldberg, *Biochemistry* 27 (1988) 3004–3011.
[44] F. Gu, Z. Xi, I.H. Goldberg, *Biochemistry* 39 (2000) 4881–4891.
[45] T. Kusakabe, M. Uesugi, Y. Sugiura, *Biochemistry* 34 (1995) 9944–9950.
[46] A. Stassinopoulos, J. Ji, X. Gao, I.H. Goldberg, *Science* 272 (1996) 1943–1946.
[47] W. Saenger, *Principles of Nucleic Acid Structure*, Springer, New York, 1984.
[48] A. Stassinopoulos, I.H. Goldberg, *Bioorg. Med. Chem.* 3 (1995) 713–721.
[49] L.S. Kappen, I.H. Goldberg, *Science* 261 (1993) 1319–1321.
[50] O.D. Hensens, G.L. Helms, D.L. Zink, D.-H. Chin, L.S. Kappen, I.H. Goldberg, *J. Am. Chem. Soc.* 115 (1993) 11030–11031.
[51] O.D. Hensens, D.H. Chin, A. Stassinopoulos, D.L. Zink, L.S. Kappen, I.H. Goldberg, *Proc. Natl. Acad. Sci. USA* 91 (1994) 4534–4538.
[52] Z. Xi, Q.K. Mao, I.H. Goldberg, *Biochemistry* 38 (1999) 4342–4354.
[53] L.S. Kappen, I.H. Goldberg, *Biochemistry* 32 (1993) 13138–13145.
[54] L.S. Kappen, I.H. Goldberg, *Biochemistry* 36 (1997) 14861–14867.
[55] L.S. Kappen, Z. Xi, I.H. Goldberg, *Bioorg. Med. Chem.* 5 (1997) 1221–1227.

[56] L.S. Kappen, I.H. Goldberg, *Biochemistry* 38 (1999) 235–242.

[57] A. Stassinopoulos, I.H. Goldberg, *Biochemistry* 34 (1995) 15359–15374.

[58] L.S. Kappen, I.H. Goldberg, *Biochemistry* 34 (1995) 5997–6002.

[59] P.C. Dedon, I.H. Goldberg, *Biochemistry* 31 (1992) 1909–1917.

[60] K.M. Weeks, D.M. Crothers, *Cell* 66 (1991) 577–588.

[61] C.F. Yang, A. Stassinopoulos, I.H. Goldberg, *Biochemistry* 34 (1995) 2267–2275.

[62] W.D. Wilson, L. Ratmeyer, M.T. Cegla, J. Spychala, D. Boykin, M. Demeunynck, J. Lhomme, G. Krishnan, D. Kennedy, *New J. Chem.* 18 (1994) 419–423.

[63] C.-C. Cheng, Y.-N. Kuo, K.-S. Chuang, C.-F. Luo, J.W. Wen, *Angew. Chem. Int. Ed.* 38 (1999) 1255–1257.

[64] Z. Xi, G.B. Jones, G. Qabaja, J. Wright, F. Johnson, I.H. Goldberg, *Org. Lett.* 1 (1999) 1375–1377.

[65] L.F. Povirk, I.H. Goldberg, *Proc. Natl. Acad. Sci. USA* 79 (1982) 369–373.

[66] L.F. Povirk, I.H. Goldberg, *Proc. Natl. Acad. Sci. USA* 82 (1985) 3182–3186.

[67] L.F. Povirk, C.W. Houlgrave, *Biochemistry* 27 (1988) 3850–3857.

[68] P. Zheng, C. Liu, Z. Xi, R.D. Smith, I.H. Goldberg, *Biochemistry* 37 (1998) 1706–1713.

Advances in DNA Sequence-specific Agents 04 (2002) 105–137

Sequence-specific DNA binding
by short peptides

Takashi Morii* and Keisuke Makino

Institute of Advanced Energy, Kyoto University, Uji, Kyoto 611–0011, Japan

1. Introduction

Gene regulation events, such as the transcriptional activation, require the assembly of multiprotein complexes at certain DNA sequences. It has become clear that transcription is controlled by multi-protein complexes in which the sequence-specificity of DNA binding by each protein is modulated by the combinatorial interactions between the proteins themselves [1–5]. Transcription factors alone do not bind to DNA with enough specificity to discriminate a unique gene from the whole genomic DNA [6–9]. Rather, it is proposed that the multi-protein complex possesses enough specificity to select a specific gene regulatory sequence from the genomic DNA. The protein–protein interactions, in addition to the protein–DNA interactions, play a fundamental role in defining the highly specific recognition of DNA by multi-protein complexes. Therefore, understanding the sequence-specific DNA binding by proteins requires a coherent elucidation of the mechanisms in which the protein–DNA interaction and protein–protein interaction complement one another to enhance the specificity of recognition.

Recent progress in structural and biochemical studies of the sequence-specific DNA binding proteins have provided useful information about the hydrogen-bonded interactions and hydrophobic contacts between amino acid residues in the proteins and the functional groups in DNA [10–13]. Unfortunately, a universal code for the recognition between proteins and nucleic acids has yet to be generalized [14]. However, these structural and biochemical studies have revealed several features of how proteins recognize specific DNA sequences: the proteins use relatively small regions to directly contact several base pairs of DNA, and most of the sequence-specific DNA binding proteins become functional only in dimeric or oligomeric forms. The first characteristic is well demonstrated by the discovery of growing numbers of DNA-binding motifs, such as helix-

*Corresponding author

E-mail address: t-morii@iae.kyoto-u.ac.jp

Advances in DNA Sequence-specific Agents, Volume 4 G. B. Jones (Editor)

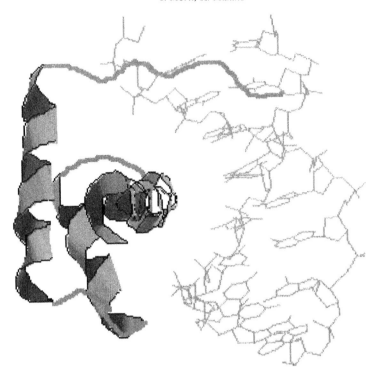

Fig. 1. Drawing of the engrailed homeodomain-DNA complex shows presentation of the recognition helix to the major groove with additional interaction between N-terminus arm and the adjacent minor groove [19]. The protein is shown as a ribbon representation, with DNA as a stick figure. The ribbon representation is generated by a program MolView (version 1.3) [149].

turn-helix [10, 11, 13, 15–18], homeo-domains (Fig. 1) [11, 13, 19–23], basic leucine zippers (Fig. 2) (bZIP) [11, 13, 24- 28], basic helix-loop-helix (Fig. 3) (bHLH) [11, 13, 29–36], and at least three types of zinc fingers [11, 13, 37–44].

These motifs tend to use α-helices in direct contact with the DNA major groove. Because most of the direct contacts between the amino acids and nucleic acid bases are made within such recognition helices, the rest of proteins can be regarded as an architecture to position the recognition helices at a proper geometry with respect to the specific DNA sequences [11]. Formation of a well-ordered dimer would determine the relative orientation of each monomer. Positioning of the recognition helices is further constrained by such a quaternary structure formation. Thus, the shape and size of the dimerization module would be critical for the final positioning of the recognition helices. In addition to these direct contacts, noncovalent and specific protein–protein interactions between the protein monomers are also important for the highly specific molecular recognition mechanisms associated with protein–DNA interactions [5, 17, 45]. Similar to the aforementioned strategy of the transcriptional control by multiprotein complexes, the sequence-specific DNA binding of homo- or heterodimers of transcription factors and bacterial

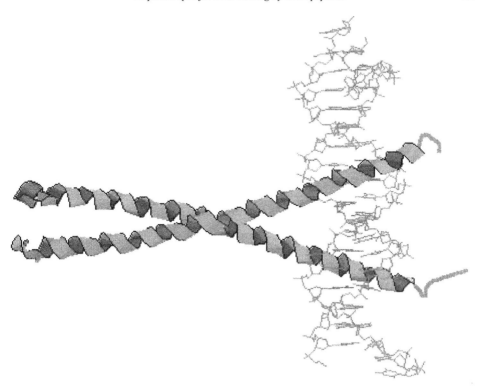

Fig. 2. Drawing of the basic leucine zipper protein GCN4-DNA complex shows presentation of the recognition helix to the major groove [79]. The protein is shown as a ribbon representation, with DNA as a stick figure [149].

repressors is also modulated by a combination of protein–protein and protein–DNA contacts [46, 47, 48]. The sequence-specific DNA binding of dimeric protein offers the simplest example for the recognition event including both the protein–DNA and the protein–protein interactions.

This review focuses on the recent development in designing novel sequence-specific DNA binding peptides by using a combination of synthetic organic, biochemical, and molecular biological approaches to study the principles of molecular recognition associated with protein–DNA interactions. Understanding recognition from the chemical and physical standpoints will require a better understanding of the energetic differences between the specific and nonspecific protein–DNA interactions. However, understanding the energetics of these protein–DNA complexes seems quite complicated because (i) recognition always involves a set of contacts and (ii) it is difficult to dissect the interactions in a way that assigns specific energetic contributions to individual contacts [11]. Because structural elements of natural proteins are less easily separated to study the role of dimer formation in modulating the affinity and cooperativity of protein–DNA interactions, many model systems have been developed to understand the functional roles of dimerization on the sequence-specific DNA binding properties of dimeric proteins. The

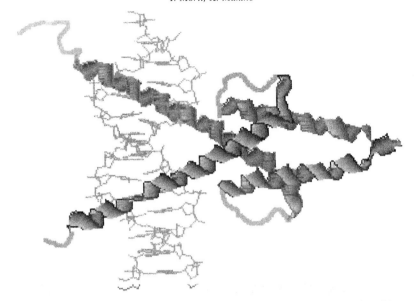

Fig. 3. Drawing of the basic helix-loop-helix protein MyoD-DNA complex shows presentation of the recognition helix to the major groove [36]. The protein is shown as a ribbon representation, with DNA as a stick figure [149].

model systems described in Section III address the issues of protein–protein and protein–DNA recognition in far greater detail than is possible with the native protein systems. These peptide dimers apply a steric constraint on the two DNA contact regions of the dimeric peptides since formation of the well-ordered dimer determines the relative orientation of each monomer. The position of the recognition helices relative to DNA is constrained by such quaternary structure formation. Moreover, this positioning would also be dictated by the shape and size of the dimerization module. Another role for the protein dimerization domain is to modulate the cooperativity of DNA binding by noncovalent protein–protein interactions. The protein–protein interaction plays an essential role in both enhancing the selectivity of specific DNA binding and in increasing the sensitivity of equilibrium binding to changes in protein concentration. Model systems with noncovalent dimerization domains are described in Sections IV, V and VI.

Design of molecules that can recognize desired DNA sequences has been one of the major challenges in the field of molecular recognition. However, few examples have been reported of novel molecules that can recognize more than five base-pair DNA sequences with reasonable selectivity. The most successful strategy to date uses polyamides containing N-methylpyrrole and N-methylimidazole amino acids to target a wide variety of sequences [49]. Because these polyamides utilize the relatively simpler information in the minor groove, as compared to that of the major groove, to recognize the DNA sequences, they suffer some diminished specificity for the A/T and T/A sequences. However, recent development of a new polyamide unit enables discrimination of these sequences [50]. The major groove is rich in its sequence-dependent information, but the triple helix approach, which is limited to the purine containing sequences, is the sole example to recognize the DNA sequences from the major groove [51]. There are numerous

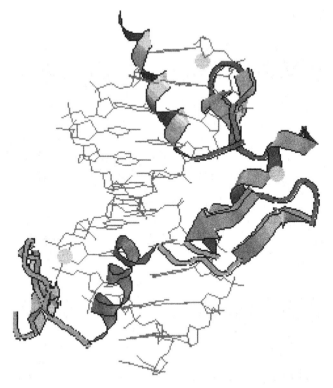

Fig. 4. Drawing of the C_2H_2 zinc finger protein Zif268-DNA complex shows three consecutive zinc finger modules fit into the major groove [40]. The protein is shown as a ribbon representation, with DNA as a stick figure [149].

alternative approaches, not reviewed here, to design sequence-specific DNA binding proteins. In particular, phage display has proven to be a powerful tool for selecting a DNA binding protein, especially the C_2H_2 zinc finger proteins (Fig. 4), that would recognize desired DNA sequences [52–56]. Due to the obvious cooperative interactions between each C_2H_2 zinc finger module, it appears necessary to perform sequential selections to optimize the affinity and sequence selectivity of the proteins [57, 58].

Recent structural and biochemical studies have provided a far greater understanding of protein–DNA interactions, but there still remains much work to be done. Why do so many proteins use α-helices for recognition? It is not yet clear whether the α-helical motif has characteristic DNA sequences that are suitable for the binding site. Is it possible to design α-helices that can fit in the major groove to recognize any given three to four base pairs? Why are disordered regions like the basic region of the leucine zipper protein used for DNA binding? Do these induced fit mechanisms provide rapid binding while retaining a set of protein–DNA contacts that are essential for specificity? Is it necessary to limit the overall binding energy and modulate appropriate binding kinetics of DNA-binding protein for gene regulation? Attempts to design novel DNA-binding proteins and/or peptides will provide an ultimate test of our understanding of protein–DNA interactions.

2. Monomeric DNA binding peptides

A. Use of the native protein motif

Most native sequence-specific DNA binding proteins bind in the major groove by using a simple secondary structure, usually an α-helix, which is complementary to the structure of the B-DNA major groove. Structural studies and sequence comparisons revealed that many DNA-binding proteins could be grouped into classes that use related structural motifs for recognition. It seems reasonable to assume that they may provide the most convenient scaffolds for the design of new DNA-binding peptides and/or proteins [11]. Unlike many of the other motifs, the HTH motif is not a separate, stable domain [59–61]. Although the second helix is often called the "recognition" helix, the outside of the HTH units can also have a significant role in recognition [62, 63]. There is no single pattern or simple code in designing DNA-binding proteins [11, 17]. The homeodomain (Fig. 1), a DNA-binding motif seen in a large family of eukaryotic regulatory proteins, contains a variant of the HTH motif [11, 64–66]. Unlike the isolated HTH unit, the homeodomain forms a stable, folded structure and can bind DNA by itself [67, 68]. However, the outside of helix3, such as the N-terminus arm that fits into the minor groove, again makes critical contacts for recognition [19, 20, 69]. It seems likely that the precise DNA-binding specificity of the homeodomain is modulated by interactions with other proteins.

Monomeric DNA-binding domains have been synthesized by the solid-phase peptide synthesis strategy. A synthetic 52-residue peptide identical to the C-terminus domain of Hin recombinase (190 amino acids) with the HTH DNA binding motif has been shown to bind to Hin recombination sites [70]. The 52-residue peptide was then converted into a sequence-specific DNA-cleaving peptide by covalent attachment of ethylenediamine-tetraacetic acid to the N-terminus of the peptide. In the presence of Fe(II), the peptide cleaved DNA at Hin recombination sites. The cleavage data revealed that the amino terminus of the 52-residue peptide is bound in the minor groove of DNA near the symmetry axis of Hin recombination sites [71]. In an effort to design sequence-specific DNA-cleaving peptides consisting wholly of naturally occurring α-amino acids, the tripeptide GGH was attached to the N-terminus of the 52-residue Hin peptide [72]. In the presence of Ni(OAc)$_2$ and monoperoxyphthalic acid, the 55-residue peptide cleaved at a single deoxyribose position on one strand of each binding site [73–75].

In order to convert a dimeric DNA-binding protein into a functional monomeric protein, stable folded monomeric variants of λ Cro, a dimeric DNA binding protein with the HTH motif, was designed by insertion of a β-hairpin sequence. The dimer interface of λCro consists of a β-strand from the C-terminus of each subunit, which form an antiparallel β-ribbon [59]. Under low salt and low temperature conditions, specific binding of the monomeric Cro to operator DNA was observed. Although binding of the monomeric Cro is reduced by at least 2000-fold compared to the wild type Cro dimer, detailed binding studies of the monomeric variants are quite useful in elucidating the role of dimerization and its connection to the cooperativity [76].

The homeodomain of *Antennapedia* was synthesized by segment condensation with protected peptide segments. The 60-residue peptide binds specifically to a TAA repeat DNA sequence in the Antp gene [77]. *De novo* design of the homeodomain has been reported by using the murine homeodomain Msx as the prototype for a "minimAl" (minimalist design and alanine scanning mutagenesis) homeodomain. The resulting protein Ala-Msx contained the consensus residues with all nonconsensus residues having been replaced with alanine. Ala-Msx bound to DNA specifically with lower affinity as compared to the parent Msx. The amino-acid sequences at the N-terminus arm of the parent Msx was required for Ala-Msx to achieve the DNA binding comparable to the parent Msx. These results indicate that the homeodomain consists of consensus residues, which are sufficient for DNA recognition, and nonconsensus residues in the N-terminus arm (Fig. 1), which are required for high affinity DNA binding. In this sense, the homeodomain consists of two DNA contacting domains, the helix-turn-helix domain and the N-terminus flexible arm [78].

To date, one of the best structurally characterized bZIP family of proteins is the yeast transcription activator GCN4 [26–28,79–81]. GCN4 is known to bind DNA as a homodimer with each basic region directly contacting the major groove of DNA (Fig. 2). The native GCN4 dimer specifically binds 5'-ATGACTCAT-3' (AP1) and 5'-ATGACGTCAT-3' (CRE) sequences in a similar affinity [82–84]. The dimerization is mediated through a coiled-coil structure at the leucine-zipper domain located at the C-terminus of the basic region [79–81, 85]. The basic region of GCN4 has disordered structure in the complex with a nonspecific DNA, but is structured to an α-helix upon binding to a specific DNA sequence [86–91]. Structure of the Jun/Fos heterodimer complexed with the AP1 sequence has also been determined [92]. The basic region peptides derived from the basic leucine zipper (bZIP) protein serve as the simplest short peptides that can target DNA sequences four to five base pairs in size [26]. However, DNA binding affinity of short peptides is inherently low and it is difficult to analyze the sequence specific DNA binding without having an anchor to increase the overall affinity of the peptide conjugate. Although a monomer of the basic region peptide has been shown to bind DNA sequence-specifically [93–96], it is necessary to form dimers of the peptide to achieve DNA binding of convincing sequence-selectivity.

B. Modification of the native protein motif

The practical problem in studying DNA binding of dimeric proteins is that the binding reactions of most dimeric proteins are highly cooperative, so that monomer–DNA intermediates are not well populated at equilibrium. It is useful to identify protein fragments or peptide models that bind as a monomer autonomously. In addition, considerations of nonspecific DNA binding has become an important issue in understanding early events in DNA binding of protein dimers [97]. The binding affinity of DNA binding proteins is largely derived from a combination of nonspecific electrostatic and hydrogen bonding interactions. Sequence discrimination is achieved through contacts made primarily by side chains of amino acids.

As models for sequence-selective DNA binding proteins, peptide conjugates of rhodium(III) complexes (Fig. 5) have been developed [98, 99]. Relatively sequence-neutral binding of the metallointercalator affords the nonspecific binding energy necessary for the stable complex formation of short peptides with DNA. The rhodium(III) complexes of phi (9,10-phenanthrenequinone) and phen' [5-(amindoglutaryl)-1,10-phenanthroline]

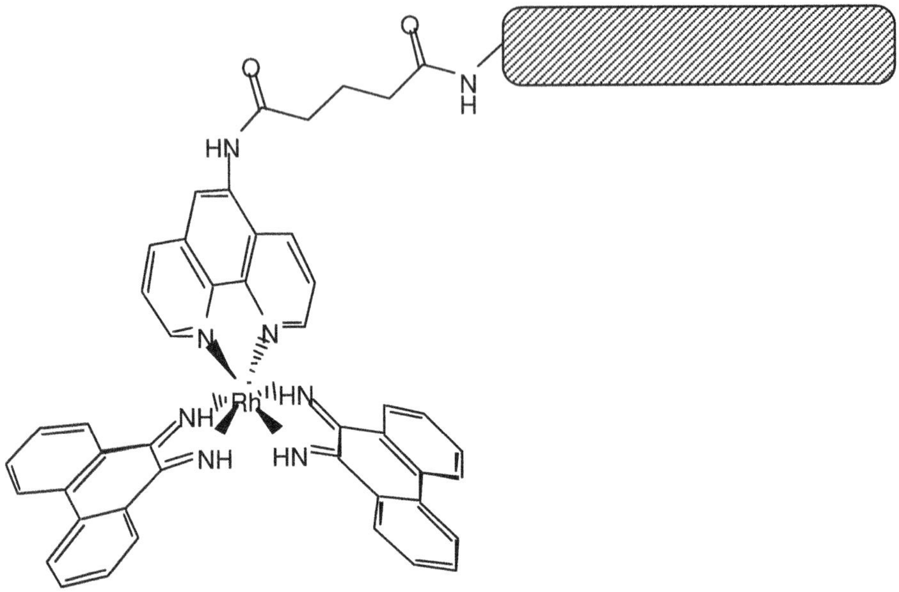

Fig. 5. Structure of the rhodium metallointercalator–peptide conjugate. The shaded box represents the peptide attached through the linker.

bound double helical DNA from the major groove and efficiently cleaved the DNA back-bone upon photoirradiation. Because the rhodium complex can deliver a peptide to the DNA recognition site, this approach offers the opportunity to explore site-specific DNA binding by small, monomeric peptides. The photocleaving reaction promoted by the rhodium (III)-phi complex permits high resolution mapping of the DNA sequences recognized by the peptide of interest.

[Rh(phi)$_2$(phen')]$^{3+}$ tethered to the oligopeptides derived from the recognition α-helix of the 434 repressor partially demonstrated the sequence selectivity of the parent protein [100]. [Rh(phi)$_2$(phen')]$^{3+}$ tethered to oligopeptides derived from the recognition α-helix of the P22 repressor afforded sequence-selective recognition of 5'-CCA-3' sequences. The highest sequence-specificity was obtained with [Rh(phi)$_2$(phen')]$^{3+}$-AANVAIAAWERAA-CONH$_2$. Glu10 in this peptide was an essential determinant of the 5-CCA-3' sequence selectivity. Mutation of Glu10 to alanine or aspartate decimated specificity. Mutation of residues other than Glu10 did not significantly alter the sequence selectivity of the peptide-metallointercalator conjugate. However, diminished specificity was observed with mutations that reduce the helical content of the peptide [101]. Interestingly, the metal-lointercalator tethered to P22 recognition α-helix did not recognize the same nucleotide sequences as P22 repressor [98]. This makes sense as the P22 repressor not only uses the recognition helix but also the outside of the helix-turn-helix motif to recognize the cognate DNA sequences. In addition, the geometry of the recognition helix in the P22 complex with DNA is severely restricted by the HTH motif and dimer configuration of the P22 repressor.

In order to increase the affinity of monomeric DNA binding peptide of bZIP, a minimalist design strategy was employed to construct a chimeric small protein containing amino acid residues derived from the GCN4 basic region [102]. Grafting of the recognition residues of the GCN4 basic region onto the avian pancreatic polypeptide, a small, well-folded protein of known structure consisting of a single α-helix stabilized by hydrophobic interaction with a type II polyproline helix, provided a small 42-residue protein. Unlike the parent avian pancreatic polypeptide, the resulting miniature protein did not fold into a stable structure under physiological conditions. Although the monomer of basic region peptide (~23-residues) recognizes the half-site of the GCN4 recognition sequence (dissociation constants of 60 nM–3 μM) in low ionic strength buffers [93, 94], the miniature 42-residue protein represents high affinity recognition of the half-site (dissociation constant of 1.5 nM) at physiological ionic strength. Addition of DNA containing the half-site of the GCN4 recognition sequence increased the helical content of the miniature protein, indicating that the induced-fit mechanism seen in the case of native GCN4 and the basic region peptides is associated with its DNA binding. This study demonstrates that binding of small domain can be improved by optimizing not only the DNA-contacting residues, but also the stability of α-helix. Unlike the case of the GCN4 basic region peptide, dimerization of two miniature proteins with a disulfide linkage provided increased affinity but a lower sequence specificity. It should be possible to create a sequence-selective dimer of the miniature protein by using a dimer interface with an appropriate configuration.

3. Covalently linked peptide dimers

A. Disulfide-bonded peptide dimers

The basic region of bZIP proteins plays critical roles in DNA binding [26]. The finding by Kim and co-workers that disulfide-bonded dimers of the GCN4 basic region peptides specifically bound the DNA sequence as observed for the native GCN4 (Fig. 6) encouraged the use of the basic region as a DNA binding unit for the design of novel DNA binding peptides. A peptide corresponding to the basic region of GCN4 (residues 222 to 252) was synthesized with a Gly-Gly-Cys linker added at the carboxyl terminus. The Gly-Gly-Cys was included to provide a flexible linker in the disulfide-bonded dimer. Gel mobility shift assays indicate that the 34-residue (PESSDPAALKRARNTEAARRSR-ARKLQRMKQ) basic region peptide dimer binds an oligonucleotide containing the GCN4 recognition element 5′-ATGACTCAT-3′ (AP1 site) with nanomolar affinity at 4°C. The DNA-binding specificities of the 34-residue basic region peptide dimer and parent GCN4 to the AP1 site were comparable as judged by the DNase I footprinting assays. However, the 34-residue basic region peptide dimer failed to bind specifically to the AP1 site at 24°C. The observed temperature sensitivity of the DNA binding by the disulfide basic region dimer suggests an additional role for the leucine zipper: to increase the thermal stability of the dimer–DNA complex. Structural studies of the basic region peptide dimer confirmed the induced fit mechanism of bZIP proteins bound to DNA in an α-helical conformation. The CD spectrum of the peptide suggests that it shows partial

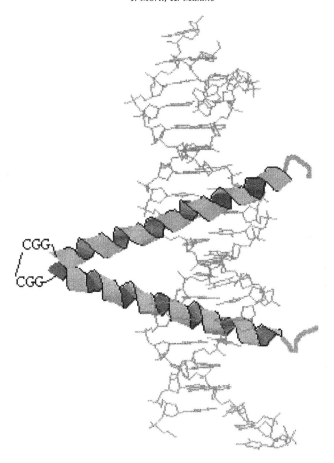

Fig. 6. A schematic representation of the binding complex between the disulfide-bonded basic region dimer and DNA. The basic region peptides in the α-helical conformation are shown as ribbon representation, with DNA as a stick figure. Coordinates for the basic region peptides and DNA are adopted from the GCN4-AP1 complex [79] and the ribbon representation is generated by the program MolView (version1.3) [149].

α-helix formation in the absence of DNA. In the presence of DNA containing the AP1 sequence, the peptide forms an α-helical structure as indicated by the substantial increase in the intensity of the helical band at 222 nm [103].

Although the 34-residue basic region peptide that contains 31 residues from the basic region of GCN4 plus the C-terminus Gly-Gly-Cys linker is a remarkably short DNA binding peptide, even shorter peptides with sequence-specific DNA-binding activity can be made. Several of the amino-terminus residues in the basic region used for the 34-residue basic region peptide have been found to be dispensable for DNA binding [104]. A series of GCN4 basic region peptides were synthesized to limit the region of GCN4 involved in sequence-specific DNA contacts. DNA footprinting experiments showed that the disulfide-bonded dimers of peptides containing as few as 20 residues of GCN4 bound DNA with sequence specificity similar to that of the intact GCN4. Peptides containing the C-terminus KLQRMKQ bound DNA with significantly greater thermal stability than peptides lacking this sequence. Further C-terminus truncations abolished the stable complex formation with DNA. Interestingly, the disulfide-bonded peptide dimers showed greater affinity for the CRE sequence 5'-ATGACGTCAT-3' compared to the AP1 sequence 5'-ATGACTCAT-3'. Because the native GCN4 dimer prefers the AP1 sequence, switching the dimerization mode from a leucine zipper coiled coil into a disulfide linkage changes sequence selectivity of the basic region dimer of GCN4. Similarly, preferential binding to the CRE sequence over the AP1 sequence was observed when the synthetic molecules, such as a bulky metal complex [105], enantiomeric templates [106], or a host-guest inclusion complex of β-cyclodextrin and the adamantyl group [107] were used as the dimerization mode for the basic region peptide of GCN4.

By extending the method of Kim and co-workers, several disulfide-bonded dimers and trimers of basic region peptides that target novel non-palindromic DNA sequences have been reported. It has been shown that an N-terminus disulfide dimer of the basic region of v-Jun binds to a palindromic DNA sequence in which an arrangement of each half site is reversed in polarity. The basic region peptide derived from v-Jun (31-residues) was modified at its N-terminus with Cys-Gly-Gly linker and dimerized through disulfide linkage. Because the parent v-Jun dimer mediated by the leucine zipper domain binds 5'-ATGACTCAT-3' sequence, the monomer of the v-Jun basic region peptide is expected to bind 5'-ATGA(C)-3' sequence. The native v-Jun monomers are dimerized at the C-terminus portion of the basic region, thus the N-terminusly dimerized v-Jun basic region peptide (v-Jun-N)$_2$ would bind to a 5'-TCAT-X-ATGA-3' sequence. Gel mobility shift assays indicated that (v-Jun-N)$_2$ bound a 5'-TCATCGATGA-3' sequence with much higher affinity than the native v-Jun dimer binding sequence at 4°C. The results indicate that individual basic region peptides of the dimer and the individual half sites of the DNA can be recombined in various sequences to form new peptide dimers that can target new DNA sequences [108].

Further extension of this strategy utilized the v-Jun derived basic region peptide with the Gly-Gly-Cys linker at either the N- or the C-terminus (v-Jun-N and v-Jun-C) [109]. These two peptides can be combined to form (v-Jun-N)$_2$, (v-Jun-C)$_2$ and (v-Jun-C/v-Jun-N) dimers. The heterodimer (v-Jun-C/v-Jun-N) was synthesized from 2-thiopyridyl-(v-Jun-C) and v-Jun-N. Gel mobility shift assays indicated that each of the three peptide dimers recognized the appropriate binding site. (v-Jun-C)$_2$ and (v-Jun-N)$_2$ bind 5'-ATGACGTCAT-3' and 5'-TCATCGATGA-3' sequences, respectively. The tandem dimer (v-Jun-C/v-Jun-N) bound the expected 5'-ATGACGATGA-3' sequence. However, DNase I footprinting for the (v-Jun-C/v-Jun-N) — DNA complex revealed incomplete protection at the expected binding sequence and partial protection on the bases flanking the binding site, which was not the case for the homodimers. These results indicate that the monomer of v-Jun basic region peptide fits the 5'-ATGAC-3' sequence and that the basic region peptide dimers bind to the appropriately rearranged DNA sequences. DNA binding of the tandem dimer is not as specific as that of the homodimers and revealed half-specific binding nature. In addition, the results suggest that a fine-tuning of the dimer linkage, such as optimizing the position of Cys, is necessary to design a peptide dimer with a specific DNA-binding activity. A similar approach has been carried out to design a trimer of the basic region peptide that can cover a 16 bps DNA sequence [110].

It is now possible to design peptide dimers that possess DNA-binding specificities different from the native bZIP by appropriately arranging two basic region peptides with an artificial dimerization domain, such as a bulky metal complex [105, 111, 112], an enantiomeric bridged biphenyl derivative [106, 113–116], a host-guest inclusion complex of β-cyclodextrin and the adamantyl group [94, 107, 117], and a lysine residue [118]. Size, shape, polarity and chirality of the dimerization domain will significantly affect the DNA binding of peptide dimers. The next section describes the use of transition metal and C_2 chiral bridged biphenyl dimerization domains to design the active quaternary structure of peptide dimers.

B. Transition metal complexes as the dimerization domain

One of the promising approaches to the assembly of protein structural domains is the template-directed synthesis of artificial proteins [119–122]. The well-defined geometries of transition metal ion complexes have been successfully used in assembling short peptides into functional forms [123–125]. A series of transition metal dimerization domains were used to alter systematically the relative orientation of the GCN4 basic region peptides (Fig. 7) [105].

The bis(terpyridyl)iron(II) complex was used in place of the GCN4 coiled coil to assemble two GCN4 basic region peptides (29-residues). The relative orientation and spacing of two basic region peptides was dictated by using terpyridyl ligands that differed in substitution pattern. Attaching the G29 peptide to three different terpyridyl ligands TS, TB and TL through disulfide-bonds and subsequent metal complexation afforded G29 dimers [G29TS]$_2$Fe, [G29TB]$_2$Fe and [G29TL]$_2$Fe, respectively. The binding of [G29T]$_2$Fe complexes to the CRE (5'-ATGACGTCAT-3') or AP1 (5'-ATGACTCAT-3') sequence was measured by gel mobility shift assay. [G29TS]$_2$Fe binds the CRE sequence as well as the native GCN4 dimer at 4°C. In contrast, neither [G29TB]$_2$Fe nor [G29TL]$_2$Fe bound CRE even at submicromolar concentrations of the dimers. GCN4 and the disulfide-linked G29 peptide dimer bind both AP1 and CRE sites with comparable affinity. Interestingly, with a bulky metal complex as a dimerization domain for the basic region peptide of GCN4, the dimeric peptide [G29TS]$_2$Fe preferentially bound the CRE sequence over the non-palindromic AP1 sequence. Because [G29TS]$_2$Fe and the disulfide-linked G29 peptide dimer binds CRE sequence with comparable affinity, the shape, size, and possibly the positive charge of the terpyridyl metal complex reduced the affinity of [G29TS]$_2$Fe to the AP1 sequence. These results clearly demonstrate that the orientation and the relative displacement with which the basic regions emerged from the dimerization domain are important in controlling the sequence selectivity of a dimer [92]. Also notable is the fact that nonspecific DNA binding of [G29TS]$_2$Fe is much higher than the simple disulfide dimer (G29)$_2$. This result is somewhat surprising because the G29 dimer with metal complex dimerization domain can discriminate CRE against AP1 by a factor of 150. Additional +2 positive charges of the metal complex dimerization domain might increase

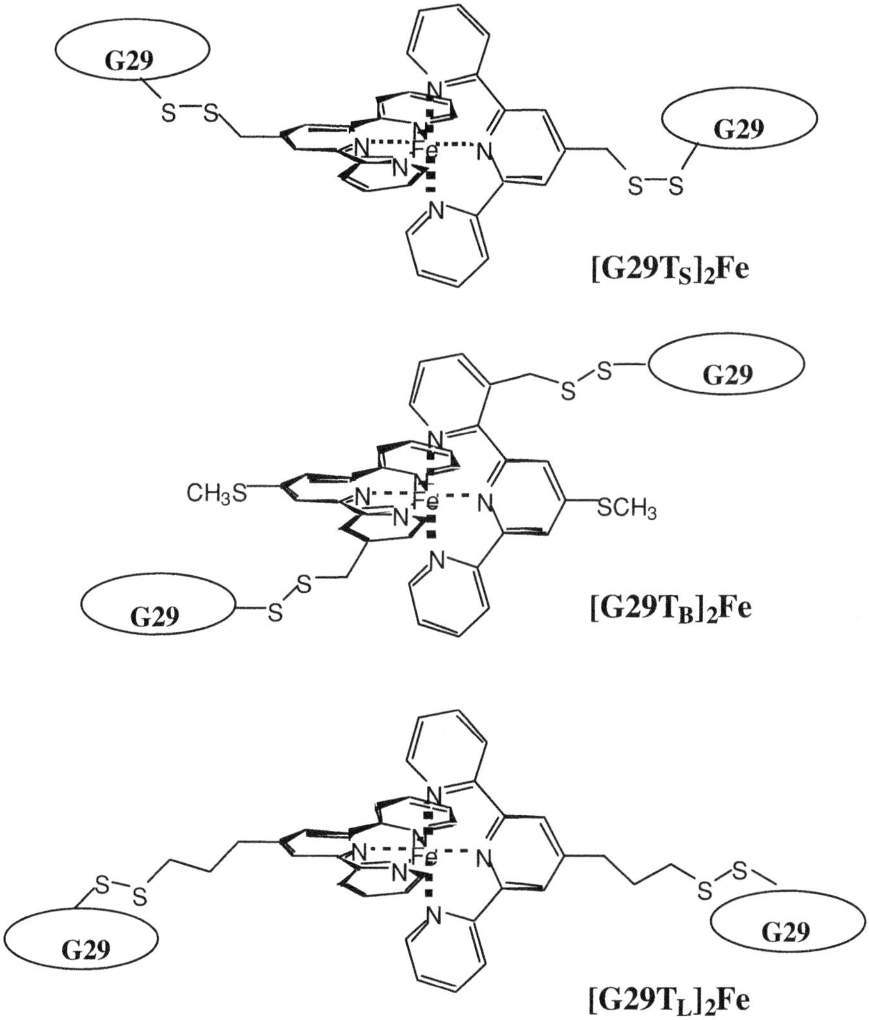

Fig. 7. Schematic representations of the metallopeptides with a series of transition metal dimerization domains. The oval represents the GCN4 basic region peptide G29.

the nonspecific binding affinity of [G29Ts]$_2$Fe [111]. Although the exact mechanism by which [G29TS]$_2$Fe discriminates CRE against AP1 is not clear, the bulky and charged metallo dimerization domain certainly acts in a steric manner or by its interaction with the C-terminus portion of the peptide [112]. Nonetheless the finding that the highly selective binding of [G29TS]$_2$Fe to CRE against AP1 is modulated by decreasing the affinity of [G29TS]$_2$Fe to AP1 is quite important for the design of DNA-binding peptide dimers of next generation.

C. C₂ chiral bridged biphenyl group as the dimerization domain

The purpose of this model is to address a question of how the three-dimensional shape of the dimerization module contributes to the sequence-specific DNA recognition by protein dimers through the arrangement of monomeric peptide motifs on a C_2 chiral template that acts as an artificial dimerization module. In the simplest case, a monomer of the direct DNA-contacting motif can be regarded as a simple α-helix. There are at least two types of constraints possible for arranging two α-helices upon formation of the dyad symmetric protein dimer (Fig. 8). One is a constraint either of the polarity that allows two orientations, in an N-terminus to N-terminus (N-N) or in a C-terminus to C-terminus (C-C) arrangement for two helices. Another one is a chiral constraint that defines a right-handed or a left-handed arrangement of two helices with respect to the C2 axis perpendicular to the major groove of DNA. Combination of these constraints results in four different types of dimers.

These four differently constrained peptide dimers have been obtained by using the enantiomers of a C_2 chiral template and tethering at either the C-terminus or N-terminus of peptides. The enantiomeric templates were synthesized from (*9R, 10R*)- and (*9S, 10S*)-*trans*-9, 10-dihydrophenanthrene-9, 10-diol as the covalently bonded dimerization module (Fig. 9) [126]. Assuming that the biphenyl chromophore is facing toward the groove of DNA, rather than exposed to a solvent, the (*9R, 10R*)-isomer would achieve a right-handed geometry of two DNA binding peptides, while the (*9S, 10S*)-isomer would achieve a left-handed geometry. The iodoacetyl group of the template was used to generate a covalent linkage between the peptide and each chiral template through a specific reaction with a SH group of unique cysteine residue in the peptides.

This approach was first employed to evaluate the DNA binding of peptide dimers derived from the basic region of bHLH protein, MyoD [113–115]. It has been demonstrated that only one of the four differently constrained peptide dimers binds specifically to a DNA sequence that is recognized by native MyoD. Specifically, both right- and left-

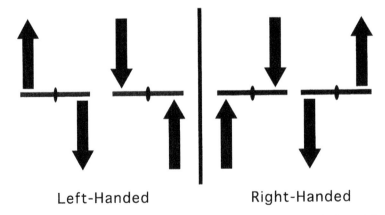

Left-Handed   Right-Handed

Fig. 8. Possible arrangements of dimers with regard to the handedness in the dimerization and the polarity of peptides. Arrows indicate the peptides with N-terminus to C-terminus polarity. Horizontal bars represent dimerization domain with dyad axis.

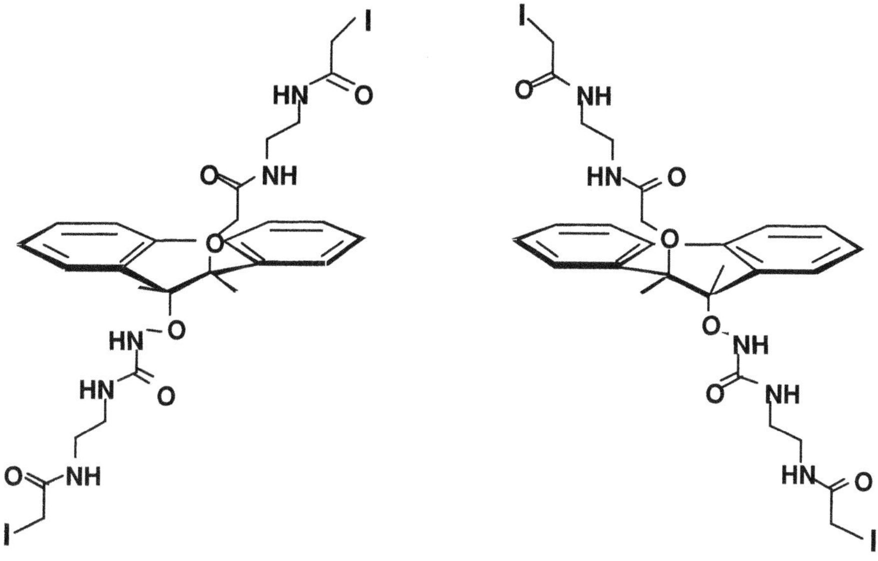

Fig. 9. Structures of the C_2 chiral dimerization domain derived from (*9R, 10R*)-*trans*-9, 10-dihydrophenanthrene-9, 10-diol (left) and (*9S, 10S*)-*trans*-9, 10-dihydrophenanthrene-9, 10-diol (right) with the iodoacetyl groups.

handed C-C dimers bound the DNA sequence recognized by the native MyoD. However, competition experiments with non-specific competitor DNA revealed that the peptide dimer with right-handed and C-C constraints bound more specifically to the native MyoD binding sequence. In fact, a report on a crystal structure of MyoD bHLH domain-DNA complex later proved that the basic region of MyoD dimer lies in the major groove of DNA in the C-C orientation and right-handed geometry (Fig. 3) [36].

The effects of geometrical constraints of the dimerization domain on the sequence-selective DNA binding were further evaluated using the basic region of the GCN4 peptide dimers that were systematically varied in the dimerization region [106]. The C-terminus dimers of the GCN4 basic region peptide (G23/C, Fig. 11) CR and CS bind 5′-ATGACTCAT-3′ (AP1) and 5′-ATGACGTCAT-3′ (CRE) sequences recognized by the native GCN4 by the same sequence-specific DNA recognition mechanisms. Apparent equilibrium dissociation constants of the peptide dimers CR and CS were almost identical for the CRE sequence (K_d = 0.26 and 0.25 nM at 4°C), and (K_d = 0.30 and 0.31 nM at 4°C). Thus, each GCN4 basic region dimer with either the right-handed (CR) or left-handed (CS) geometry showed almost identical affinity to the native GCN4 binding sequences as the disulfide-bonded basic region peptide dimers. In addition, both CR and CS dimers exhibited the characteristic increase in the intensity of the CD signal at 222 nm (helical band) in the presence of specific DNA sequences.

In the case of MyoD basic region dimers with the same chiral templates, the right-handed dimer showed slightly higher affinity to the native MyoD binding sequence [115]. Possible explanations for the reason why these two systems yielded different results come from an inspection of the X-ray crystal structures of GCN4 and MyoD bound to their

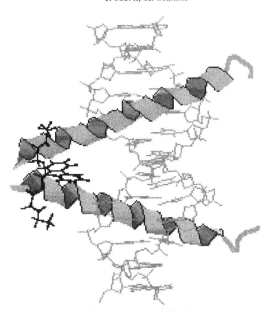

Fig. 10. A schematic representation of the binding complex between the basic region dimer with chiral dimerization domain and DNA. Coordinates for the basic region peptides and DNA are adopted from the GCN4-AP1 complex [79] and the ribbon representation is generated by the program MolView (version1.3) [149].

cognate DNA sequences (Figs 2 and 3) [36, 79–81]. Both bHLH and bZIP proteins juxtapose the α-helices rich in basic amino acid residues into the DNA major groove. The C-terminal portions of the two basic regions of the GCN4 dimer are positioned next to each other due to the structural constraint induced by the leucine zipper coiled-coil. On the other hand, the C-terminal portions of the two basic regions are separated further in the MyoD dimer [36]. Such differences in the arrangements of DNA contacting regions in the parent protein–DNA complexes could yield different results for the MyoD system and GCN4 system. It has been demonstrated that six amino acid residues in between the basic region and the leucine zipper, termed the spacer segment, contribute to the CRE/AP1 selectivity [83, 84, 127–129]. The GCN4 basic region peptide used here contains only the first three residues of the spacer segment. Thus, the basic region itself and structure of the enantiomeric template would be responsible for the observed selectivity of CR and CS dimers to the CRE and AP1 sequences.

Peptide dimers of the basic leucine zipper protein with non-native monomer arrangements, an N-terminus to N-terminus orientation, were synthesized by using the C_2 chiral templates described above [116]. The amino acid sequences of oligopeptides, C/G23 and C6/G23 are derived from that of the basic region of yeast transcription factor GCN4 (Fig. 11). NS and NR are constrained at the N-terminus of C/G23 peptide, while N6R and N6S are at the 6th position of C6/G23. Two DNA sequences RCRE (5'-TCATC-GATGA-3') and R2CRE (5'-GTCATATGAC-3') were designed to possess a reversed half-site orientation [105] as compared to the native GCN4 binding sequence CRE (5'-ATGACGTCAT-3'). In the RCRE sequence, the central CG sequence of the CRE remains

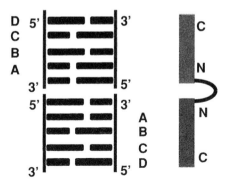

5'-GTCAT-ATGAC-3' : RCRE   CPAALKRARNTEAARRSRARKLQ : C/G23

5'-TCATC-GATGA-3' : R2CRE   DPAALCRARNTEAARRSRARKLQ : C6/G23

5'-ATGAC-GTCAT-3' : CRE   DPAALKRARNTEAARRSRARKLC : G23/C

Fig. 11. Amino acid sequences of C/G23, C6G23 and G23/C peptides are shown with the target DNA sequences of N-terminus dimers (RCRE and R2CRE) and C-terminus dimer (CRE).

unchanged, while orientation of the outer 4-bp sequences is reversed. In the case of R2CRE, orientation of the whole half-site of CRE, 5'-ATGAC-3', is reversed. The 5'-ATGA-3' is regarded as a half-site and the central CG step as a spacer sequence for the RCRE, and 5'-ATGAC-3' is regarded as a half-site for the R2CRE sequence. N6R and N6S bound the R2CRE sequence with dissociation constants of 0.15 nM and 0.10 nM, respectively. This contrasts with dimers that are constrained at the N-terminus that failed to recognize the reversed sequences. Circular dichroism spectral analyses revealed that dimers constrained at the 6th position bind in the helical conformation to the reversed palindromic sequences, whereas the dimers constrained at the N-terminus bound the same sequence with lesser helical contents. One surprising effect of the chiral templates is a marked reduction in the affinity of the peptide dimer to nonspecific sequences. Affinities of N6R to the noncognate sequences, especially to the "half-matched" sequences, such as AP1 and CRE, differ considerably from that of N6S. The N6R-CRE complex is more stable than the N6S-CRE complex by a factor of 25. The N6R-AP1 complex is more stable than the corresponding N6S complex by a factor of 11. Similarly, N6R forms an 8 times more stable complex with the DNA sequence containing only a half-site (HS: 5'-ATGAC-3') than N6S does. In the case of non-relating sequence, N6R complex is more stable than the N6S complex by a factor of three. Since N6R and N6S would form the half-specific complexes with CRE, AP1 and HS, the geometrical constraints by SDHP lowers the stabilities of half-specific binding complexes of N6S. Such differences in the stabilities of non-specific binding complexes that are affected by the particular enantiomer of the template have also been observed for the MyoD derived peptide dimers [115]. Although the mechanism for this anti-selectivity is yet to be established, the chiral template of N6S reduces the affinity of the peptide dimer for the non-targeting sequence, rather than increasing the affinity for the specific sequences, possibly by interacting with the N-terminus portions of the peptides. Such an effect in the synthetic dimerization module has also been suggested for the DNA binding of metallo-peptides as described in the previous section [105]. Since the bZIP basic region dimers already have high affinities for their respective target DNA sequences, the strategy of reducing the nonspecific binding would aid in the design of the second generation homo- and heterodimers that recognize a variety of DNA sequences with high selectivity.

4. Noncovalent peptide dimers

A. Dimerization promoted by a host–guest inclusion complex

The noncovalent interaction between two monomers is another important feature in the sequence-specific DNA binding by protein dimers. The equilibrium governing the formation of dimers should be important in enhancing the selectivity of DNA binding and in increasing the sensitivity of DNA binding to the change in protein concentrations. In fact, many transcription factors form the homodimers to recognize the palindromic DNA sequences with each monomer of the homodimer fitting to each half-site. In order to understand the role of noncovalent dimer formation in modulating the affinity and cooperativity of protein–DNA interactions, DNA-binding oligopeptides capable of forming a functional quaternary structure by noncovalent interactions were designed (Fig. 12).

A host–guest inclusion complex of β-cyclodextrin (Cd) and adamantane (Ad) provides a new method to associate oligopeptides in an aqueous solution. This module in essence mimics the specific protein–protein interactions that govern the formation of homo- and heterodimers. A peptide modified with β-cyclodextrin and another peptide modified with an adamantyl group formed a noncovalent dimer that was mediated by formation of the β-Cd/Ad inclusion complex [107]. In the case of native protein dimers, reducing the dimerization ability by using the deletion mutants may complicate the situation [130–132]. It is difficult to measure the dimerization ability directly and these mutations may give an undesired effect on the folding of DNA binding domain. In contrast, the β-Cd/Ad dimerization domain is modular and gives rise to little conformational change in the process of an inclusion complex formation. In addition, β-Cd forms specific 1:1 complexes with many guest molecules in an aqueous solution [133–136]. The stability of β-Cd/guest dimerization domain can be independently controlled by changing the guest molecules without greatly affecting the DNA-binding ability of the peptide itself.

The basic region peptide derived from GCN4 has been used to test the validity of the β-Cd/guest dimerization domain. The GCN4 peptide modified at the C-terminus with an adamantyl group indeed dimerized with another peptide that had the β-cyclodextrin attached to the C-terminus, and the peptide dimer further showed specific DNA binding to the native GCN4 site. DNA binding of G23Ad and G23Cd was studied by gel mobility shift assay. The GCN4 basic region peptide with adamantyl group, G23Ad, at the concentrations below 100 nM did not show any detectable binding to the native GCN4 binding sequence (CRE: 5′-ATGACGTCAT-3′) and poly(dI-dC). In contrast, binding mixtures

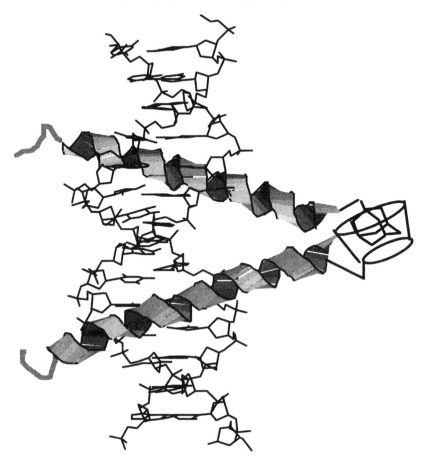

Fig. 12. A schematic representation showing the G23Ad/G23Cd dimer bound at the AP1 sequence. Gray ribbons represent the basic region peptides in the helical conformation. Shown at the C-terminus of the helices are β-cyclodextrin and the adamantyl group. Coordinates for the basic region peptides and DNA are adopted from the GCN4-AP1 complex [79] and the ribbon representation is generated by the program MolView (version1.3) [149].

containing a 1:1 mixture of G23Ad and G23Cd showed gradual appearance of a complex with lower electrophoretic mobility than CRE. These results indicate that G23Ad and G23Cd bind to DNA as a dimer. Sequence-specific DNA binding of the mixture of G23Ad and G23Cd was also confirmed by the DNase I footprinting technique. G23Ad alone did not show any detectable binding to the CRE in a dimeric form. In contrast, the mixture of G23Ad and G23Cd showed DNA binding to the CRE in a dimeric form. The increase in intensity of the CD signal at 222 nm on addition of CRE to the mixture of G23Ad and G23Cd indicates that the G23 in G23Ad/G23Cd — DNA complex adopts a helical structure as observed for the basic region of native GCN4. In order to confirm that the dimerization of G23Ad and G23Cd is attributable to the formation of an inclusion complex between β-Cd and the adamantyl group, either free β-Cd or 1-adamantanemethylamine was added to the binding mixture containing G23Ad/G23Cd and CRE. The formation of a mobility-shifted band by the 1:1 mixture of G23Ad/G23Cd was inhibited on addition of free β-Cd or 1-adamantanemethylamine to the binding mixture. These results strongly indicate that G23Ad and G23Cd form a dimer mediated by formation of an inclusion complex between β-Cd and adamantyl group.

B. Sequence-specific DNA binding by peptide heterodimers

The heterodimerization of transcriptional factors has been shown to play an important role in the specific gene control events [1]. The heterodimer formation is a consequence of the specific protein–protein interaction between two different monomers. Thus, a prerequisite for DNA binding of the heterodimer is a module that appropriately controls such specific protein–protein interactions. However, a more tantalizing question is whether or not the heterodimers actually bind non-palindromic DNA sequences. To this end, the β-Cd/guest synthetic dimerization was utilized (Fig. 12) to regulate the specific formation of heterodimer that is difficult to control in the case with native proteins [117]. Because it is the host–guest inclusion complex that regulates the dimerization [107], attachingβ-Cd to one peptide and a guest molecule to another peptide will allow a specific formation of peptide heterodimer. Upon heterodimerization through the β-Cd/adaman-tane dimerization module, the resulting heterodimer recognizes a non-palindromic DNA sequence that consists of two distinct half-sites which correspond to the native protein binding sequences.

The DNA binding regions of two different basic leucine zipper proteins, the yeast tran-scriptional activator GCN4 (G23) and an enhancer binding protein C/EBP (C23) [28], have been successfully used to design heterodimers that can recognize non-palindromic DNA sequences. Modification of the C-terminus cysteine of the peptides with mono 6-deoxy-6-iodo-β-cyclodextrin (Cd) or *N*-bromoacetyl-1-adamantanemethylamine (Ad) afforded four different peptides (G23Ad, G23Cd, C23Ad, and C23Cd) that are capable of forming specific homo- (G23Ad/G23Cd and C23Ad/C23Cd) and heterodimers (C23Ad/G23Cd and C23Cd/G23Ad). GCN4 and C/EBP are known to recognize palin-dromic sequences with a half-site of 5'-ATGAC-3' (CRE) and 5'-ATTGC-3' (CE), respec-tively [137]. Combination of these half-sites gives a non-palindromic sequence (CE/CR), 5'-ATGACGCAAT-3', as the target of the peptide heterodimer. Binding selectivity of the homo- and heterodimers to the palindromic and non-palindromic DNA sequences was compared by titration of the gel-shift. The heterodimer C23Cd/G23Ad preferentially bound to the CE/CR sequence over the palindromic CRE and CE sequences. Homodimers G23Ad/G23Cd and C23Ad/C23Cd bound to the palindromic CRE and CE sequences, respectively. However, both homodimers were found to bind the non-palindromic CE/CR sequence as well, and notably, C23Ad/C23Cd showed comparable affinity to the CE/CR and CE sequences. These results together with the partial footprinting obtained for the G23 homodimer at CE/CR sequence indicate an existence of half-matched binding complexes for DNA binding of the homo- and heterodimers. The helical band intensity of the heterodimer in the presence of the palindromic sequence (CRE or CE) is about half of that observed for the CE/CR sequence. This is also the case with the C23Ad/C23Cd in the presence of CECR21. Thus, the structure of the half-matched binding complex is partially helical and, most likely, only the monomer at the matched half-site is in the α-helical conformation.

These model studies demonstrate that discrimination between the palindromic DNA sequences by the homodimers is much more efficient than that of non-palindromic sequences by the homo- or heterodimers. A half-matched binding complex formed between a palindromic DNA sequence and a heterodimer, or between a non-palindromic DNA sequence and a homodimer, displays a structure different from that of a matched binding complex. Such subtle structural differences in the native DNA–protein dimer complexes would be critical for the subsequent specific protein–protein interactions that control the gene activation events.

C. Dimerization promoted by short peptides

The structure-based method for linking DNA-binding molecules provides powerful design strategy. Usually, two crystal structures of DNA–protein complexes are superimposed by aligning phosphates in various registers. Based on a favorable alignment of two (or more) complexes, the domains of interest can be fused directly or with linker region in between. By using this strategy, zinc fingers have been fused to a homeodomain [138, 139], or to the TATA-binding protein [140]. In order to create a dimeric zinc finger protein, computer modeling was used to design a novel dimeric protein consisting of zinc fingers 1 and 2 from Zif 268 [40], which targets a 5′-CGCCCA-3′ subsite, and the dimerization domain of GAL4 [141]. The GAL4 dimerization motif contains a coiled-coil motif that can form a homodimer or a heterodimer [142]. The resulting chimeric protein thus designed binds the predicted DNA sequence, 5′-<u>CGCCCA</u>GAGGACAGTCCTC-<u>TGGGCG</u>-3′ as a homodimer with an equilibrium dissociation constant of 7.8×10^{-19} M^2. The adaptability of this design is demonstrated by changing the Zif268 derived zinc fingers into the novel zinc fingers obtained through phage display for binding to the subsite 5′-AAGGGT-3′ [57]. The sequence specificity of the chimeric protein is primarily determined by the zinc fingers, allowing substitution of fingers to target new sites and to generate heterodimers capable of binding non-palindromic DNA sequences.

An alternative strategy to create a noncovalent dimer is the selection of dimerization elements from libraries of random peptides. Phage display was used to select and optimize 15-residue peptides that mediate dimerization of DNA-binding zinc fingers [143]. Random 15-residue peptides were fused to the N-termini of zinc fingers 1 and 2 from Zif 268, and these peptide-zinc fingers were displayed on filamentous bacteriophage. Selection against a target DNA containing an inverted repeat of the zinc fingers' binding site enriches phage that display peptides with dimeric binding activity. After seven cycles of initial selection and amplification, the phage pool was sequenced and resulted in 6 different 15-residue peptides. Each of the zinc fingers with selected peptide actually binds the target DNA sequence as a dimer. Reoptimization of the selected peptide sequences was performed in a sequential selection and amplification because the size of the starting phage pool (10^8) is far smaller than the all possible 15-residue peptide sequences (10^9). Two of the selected 15-residue peptide sequences were divided into three five-residue blocks. The block next to the N-termini of the zinc fingers is completely randomized, and then selected for sequences with higher affinity for the symmetric DNA site. The second and third blocks were randomized and reselected in subsequent reoptimization steps. These reoptimization steps resulted in consensus peptide sequences of HPMNN-LLNYV-VPKLR, and PAWLT-EYLES-MRKWR that did not show any significant similarity to the known natural proteins. The zinc fingers with these reoptimized dimerization elements bind the target palindromic DNA sequences with three order higher affinity than the initially selected zinc fingers. It is noteworthy to mention that several peptide sequences obtained during the selection procedure appeared to form high-order oligomers. Investigating the structure of the dimerization element should yield basic insights into how dimerization can be achieved with short peptides.

5. Cooperative DNA binding

A. Cooperative DNA binding by peptide dimers

The protein–protein interactions controlling the dimerization are considered to play a predominant role in the cooperative DNA binding of proteins. Because the stability of the β-Cd/guest dimerization domain can be independently controlled by changing the guest molecules without greatly affecting the DNA-binding ability of the peptide itself, the basic region peptide with the Cd/guest dimerization domain affords a useful model system to study the cooperative DNA binding of proteins that only differ in the dimerization ability. The sequence-specific DNA binding and cooperative DNA binding of peptide dimers will be further studied by using peptides with a variety of guest molecules attached at various locations.

This section describes how the stability of the dimer affects the cooperative formation of the peptide dimer–DNA complex [144]. The GCN4 basic region peptides with five different guest molecules were synthesized (Fig. 13) and their equilibrium dissociation constants with a peptide possessing β-cyclodextrin were determined. These values, ranging from 1.3 to 15 μM, were used to estimate the stability of the complexes between the dimers with various guest/cyclodextrin dimerization domains and GCN4 target sequences (Table 1). An efficient cooperative formation of the dimer complexes at the GCN4 binding sequence was observed when the adamantyl group was replaced with the norbornyl or noradamantyl group, but not with the cyclohexyl group which formed a β-cyclodextrin complex with an order of magnitude lower stability than the adamantyl group. Thus, cooperative formation of the stable dimer–DNA complex appeared to be

Table 1
Equilibrium dissociation constants K_{d1} (dimerization) and K_{d2} for the G23Ad/G23Cd, G23NA/G23Cd, G23Nb/G23Cd, G23NrA/G23Cd and G23Ch/G23Cd complexes with the AP sequence

Guest peptide	K_{d1} (M)	K_{d2} (M)	ΔG (kcal/mol^{-1})
G23Ad	$1.34 \pm 0.25 \times 10^{-6}$	$2.32 \pm 0.06 \times 10^{-11}$	-20.9
G23NAd	$1.32 \pm 0.19 \times 10^{-6}$	$1.22 \pm 0.06 \times 10^{-11}$	-21.3
G23Nb	$3.09 \pm 0.47 \times 10^{-6}$	$6.93 \pm 0.59 \times 10^{-12}$	-21.1
G23NrA	$2.68 \pm 0.42 \times 10^{-6}$	$4.31 \pm 0.47 \times 10^{-12}$	-21.5
G23Ch	$1.47 \pm 0.33 \times 10^{-5}$	$8.48 \pm 2.77 \times 10^{-10}$	-17.6

G23: Ac-DPAALKRARNTEAARRSRARKLQX-NH₂

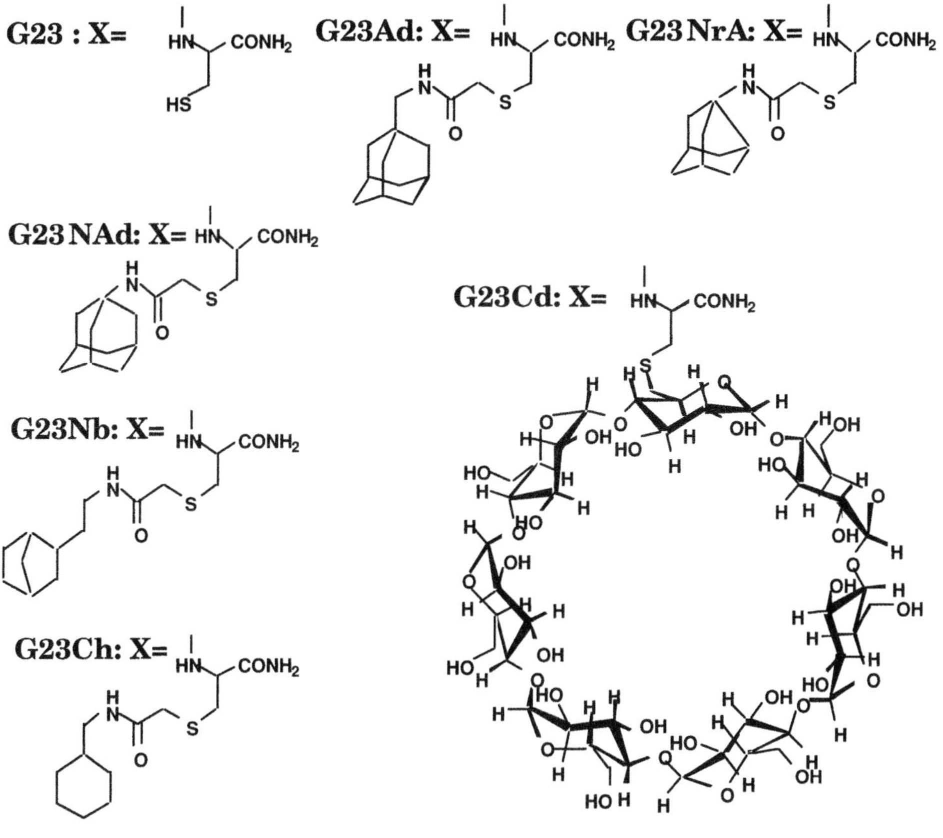

Fig. 13. An amino acid sequence for the G23 peptide and structures of the modified C-terminus Cys residue (X). The cysteine residues were modified with 6-deoxy-6-iodo-β-cyclodextrin for G23Cd and with bromoacetyl derivatives of the guest molecules for G23Ad, G23NAd,G23NrA, G23Nb and G23Ch.

effected by the stability of dimerization domain. For the peptides that cooperatively formed dimer–DNA complexes, there was no linear correlation between the stability of the inclusion complex and that of the dimer–DNA complex.

With the β-cyclodextrin/adamantane dimerization domain, the basic region peptide dimer bound preferentially to a palindromic 5′-ATGACGTCAT-3′ (CRE) sequence over the sequence lacking the central GC base pair (AP1) and that with an additional GC base pair in the middle (5′-ATGACGGTCAT-3′). Changing the adamantyl group into a norbornyl group did not alter the preferential binding of the basic region peptide dimer to the palindromic sequence, but slightly affected the selectivity of the dimer to other non-palindromic sequences. For these dimer–DNA complexes, the helical contents of the peptides in the DNA bound form were decreased with decreasing stability of the dimer–DNA complex, possibly caused by deformation of the helical structure proximal to the dimerization domain. These results indicate that the stability of the inclusion complex dominantly controls the cooperative formation of the dimer–DNA complex while the shape and size of the dimerization domain affect the sequence selectivity of peptide dimers.

The minimal energetic requirement for the cooperative formation of dimer-specific DNA complex should be highly correlated with the binding free energy and the structure of host–guest complex. However, the cooperative formation of the sequence-specific dimer–DNA complex would also depend on the inherent DNA-binding affinity and specificity of a monomeric basic region peptide. Thus, it is important to compare the cooperative interaction energy for these GCN4 basic region dimers with that for other basic region peptide dimers with the same dimerization domain.

B. Cooperative DNA binding by peptide oligomers

As described in the previous sections, both homo- and heterodimers consisting of an adamantyl peptide and a β-cyclodextrin-peptide can target the palindromic and/or non-palindromic DNA sequences. This strategy has been extended to a cooperative DNA binding by peptide homo-oligomers of an oligopeptide derived from the basic region of GCN4 [94, 145].

A series of short peptides derived from the basic region of the basic leucine-zipper protein GCN4 were synthesized to study the cooperative DNA binding to direct-repeat sequences (Fig. 14). A modified lysine residue bearing an adamantyl group at the ε-amino group was incorporated at the N-terminus position and β-cyclodextrin was attached at the C-terminus cysteine residue of the parent basic region peptide. The resulting GAdCd

G1 : Ac–XPAALKRARNTEAARRSRARKLQC–NH₂

G2 : Ac–DXAALKRARNTEAARRSRARKLQC–NH₂

G3 : Ac–XDPAALKRARNTEAARRSRARKLQC–NH₂

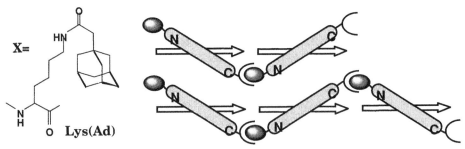

Fig. 14. Schematic representations showing the G2AdCd homo-oligomer bound on the direct-repeat sequence of CRE half-site. Hatched bars represent the basic region peptide. N and C indicate N-terminus and C-terminus of the peptide, respectively. Half-circles at the C-terminus and filled ovals at the N-terminus represent β-cyclodextrin and the adamantyl group, respectively. White arrows denote the CRE half-site. Also shown is a structure of Fmoc-Lys(Ad).

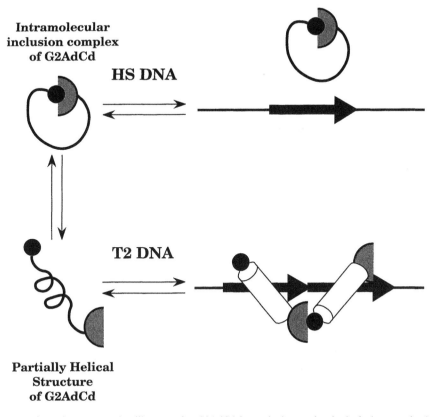

Fig. 15. A schematic representation illustrates that G2AdCd forms the intramolecular inclusion complex in the absence of specific DNA sequence and forms an α-helical trimer–DNA complex with T2 DNA. The intramolecular inclusion complex remains stable in the presence of HS DNA. Half-circles and filled ovals represent β-cyclodextrin and the adamantyl group, respectively. Filled arrows denote the CRE half-site. The cylinders represent the helical form of G2AdCd.

peptide possesses both host and guest molecules in the same peptide chain. The peptide without β-Cd (GAd) bound direct-repeat sequences of a half-site of the native GCN4 binding site without any cooperativity. DNA binding of the GAdCd peptides to single, double and triple direct-repeat of the CRE half-site were compared by titration of the gel-shift. Interestingly, the GAdCd peptide did not bind the single CRE half-site (5'-ATGAC-3'), although a peptide lacking the β-cyclodextrin group formed a specific monomer–half-site complex. GAdCd bound the double direct-repeat sequence (T2: 5'-ATGAC-ATGAC-3') as a dimer in a cooperative manner. Moreover, cooperative formation of a 3:1 GAdCd-DNA complex was observed for a triple direct-repeat sequence with no monomer–DNA complex of GAdCd being observed for the double or triple direct-repeat sequence. In the absence of DNA, GAdCd was found to form an intramolecular host–guest complex (Fig. 15). Formation of this cyclic peptide appears to reduce the affinity of monomeric GAdCd to the CRE half-site as compared to that of GAd. Thus, the highly selective binding of GAdCd observed here was accomplished by (i) its cooperative nature of DNA binding and by (ii) destabilization of its non-specific DNA binding complex.

The synthetic DNA-binding peptides with self-oligomerization ability bind with positive cooperativity to DNA sequences with multiple direct-repeats. Introduction of an explicit interaction between the peptides, the host–guest interaction in the present study, onto the DNA-binding ability results in increased selectivity to the target DNA sequence. This is evident from the reduced affinity of GAdCd to the HS sequence (5'-ATGAC-3') as compared to that of GAd. In this system, stability of the cyclic peptide formed via intramolecular inclusion complex is another important factor in reducing the affinity of GAdCd peptides to an isolated half-site as illustrated in Figure 15. The balance of intramolecular versus intermolecular interactions accounts for binding selectivity.

6. Factors governing the sequence selectivity of peptide dimers

DNA-binding proteins generally consist of more than two DNA contacting regions which ensure the selectivity of recognition [11, 13]. A class of proteins binds DNA with multiple DNA-binding modules connected through covalent linkages, and the other with the noncovalent formation of homo- and heterodimers. In the former class of proteins, the greatest progress to date has been achieved in studying the sequence-specific DNA binding of the C_2H_2 zinc finger proteins. Phage display techniques are successfully used to design novel three-finger proteins with desired specificities [57]. In addition, connecting two three-finger peptides with an appropriately designed linker has provided six-finger proteins with stability and sequence selectivity higher than the three-finger protein [146, 147]. In the later class of proteins, the protein–protein interactions controlling the dimerization are considered to play significant roles in enhancing the selectivity of specific DNA binding and increasing the sensitivity of equilibrium binding to the change in protein concentrations [5, 11, 45, 148]. Despite the current interests in exploring the advantages of these covalent and noncovalent linkages on the sequence-specific DNA binding, it is difficult to compare the DNA binding of covalently linked multi-modular proteins and those of noncovalent dimeric proteins directly because the DNA contacting regions of these native proteins are different from each other.

The systems for assembling the covalent and noncovalent dimers will complement each other to ask a question of whether the cooperative binding enhances the sequence-selectivity of peptide dimers. In this section, DNA binding of peptide dimers with covalent and noncovalent dimerization domains sharing the same DNA contacting regions have been compared [145] (Fig. 16). Three sets of peptide dimers that could target the same DNA sequence, but differed in the dimerization domains were designed. A 23-aminoacids peptide derived from the basic region of yeast bZIP protein GCN4 was used as a DNA-binding domain for all of the peptide dimers. The noncovalent homodimer GAdCd possessed both adamantyl group (Ad) and β-cyclodextrin (Cd) within the same peptide chain [94]. Because the noncovalent homodimerization was regulated by the N-terminus Ad and the C-terminus Cd, GAdCd bound to the target DNA sequence with a head-to-tail dimer configuration. The same head-to-tail configuration of a noncovalent heterodimer

G2AdCd: Ac-D**X**AALKRARNTEAARRSRARKLQ**Z**-NH$_2$
X = Lys(Ad),   **Z** = Cys(Cd)

G2Ad: Ac-D**X**AALKRARNTEAARRSRARKLQ**Z**-NH$_2$
X = Lys(Ad),   **Z** = Cys(acetamide)

GCd: Ac-D**X**AALKRARNTEAARRSRARKLQ**Z**-NH$_2$
X = Pro,   **Z** = Cys(Cd)

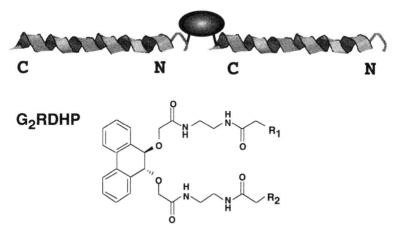

G$_2$RDHP

R$_1$ = Ac-DPAALKRARNTEAARRSRARKLQ**C**-NH$_2$
R$_2$ = Ac-**C**PAALKRARNTEAARRSRARKLQ-NH$_2$

Fig. 16. Structures and amino acid sequences for the basic region peptide dimers with covalent (G$_2$RDHP) and noncovalent (G2Ad/GCd, G2AdCd) dimerization domains.

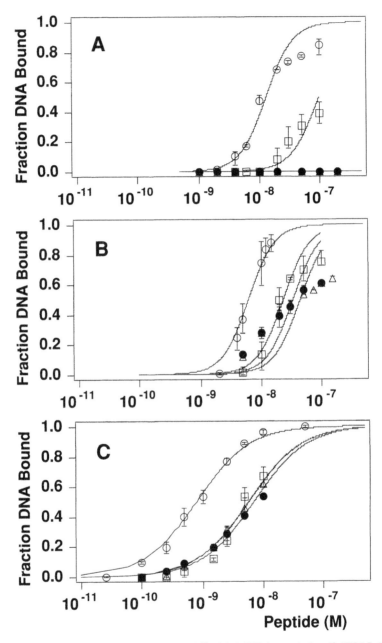

Fig. 17. Semilogarithmic plots showing the fractions of ^{32}P-labeled T2 (open circles: 5′-ATGAC-ATGAC-3′), G3A (filled circles: 5′-ATGACATAAC-3′), HS (open triangles: 5′-ATGACACTGC-3′), and T2S (open squares: 5′-ATGACCATGAC-3′) bound to the G2AdCd (A) G2Ad/GCd (B) and G$_2$RDHP (C) dimers as a function of peptide concentration. The solid curves represent the best fit for the titration data.

was achieved by using two basic region peptides: one with the N-terminus adamantyl group (GAd) and another with the C-terminus Cd (GCd). A covalent dimer G_2RDHP had a bridged biphenyl derivative [106, 115, 116, 126] as a dimerization domain that connected the C-terminus of a basic region peptide to the N-terminus of another peptide to orient two peptides in the head-to-tail configuration. Both noncovalent dimers (GAdCd and GAd/GCd) formed the specific DNA complexes within narrower ranges of peptide concentrations and showed higher sequence selectivity than the covalent dimer G_2RDHP did (Fig. 17). Among the three dimers, the noncovalent homodimer GAdCd that could form an intramolecular inclusion complex showed the highest sequence selectivity to discriminate a single base-pair mutation within the target sequence (5′-ATGAC-ATGAC-3′).

Cooperativity in the DNA binding thus plays significant roles in (i) increasing the sensitivity of equilibrium binding to the change in protein concentrations and (ii) enhancing the selectivity of specific DNA binding. The noncovalent peptide dimers GAdCd and GAd/GCd clearly exhibited the former feature. This characteristic also participated in increasing selectivity of the noncovalent dimers at the peptide concentrations required for saturating the specific DNA sequence. The noncovalent dimer GAdCd shows dramatically high sequence selectivity. However, there are not great differences in the sequence selectivity between GAd/Cd and G_2RDHP. An equilibrium involving a conformational transition of a monomeric peptide (Fig. 15) effectively reduced the stability of its non-specific binding complex hence increasing the efficacy of cooperative dimer formation at the specific DNA sequence. Thus, cooperative formation of the dimer–DNA complex through noncovalent interactions of monomers does not necessarily warrant the high sequence selectivity. The affinity of monomer to nonspecific sites and the structure of dimerization domain, which could modulate the cooperative interaction energies, are important factors to increase the selectivity in the cooperative DNA-binding systems. The mechanism involving an equilibrium that leads to a formation of less active DNA-binding species, such as a conformationally altered monomer, would be quite important for increasing the selectivity and for exerting the cooperative nature of the DNA-protein complex formation.

7. Concluding remarks

Future research will continue to use model systems which will allow a greater detailed investigation of the thermodynamic and kinetic aspects of sequence-specific recognition of DNA by protein dimers than is possible with the native systems. The results obtained from the covalent and noncovalent peptide dimers described here will complement each other in understanding the principle roles of protein–protein interaction in the sequence-specific recognition of DNA by protein dimers.

It is noteworthy to mention that the observed high selectivity of synthetic peptide dimers resulted from the reduced affinity of peptides to the nonspecific DNA sequences. By appropriately utilizing the mechanisms by which peptide dimers discriminate DNA sequences, it should be possible to design a novel DNA binding peptide that can target a specific DNA sequence with high selectivity even under physiological conditions. Studies using the enantiomeric dimerization domains have demonstrated that subtle structural change in the dimerization domain causes destabilization of nonspecific peptide–DNA complexes. The second generation synthetic dimerization domain will be designed to increase the sequence selectivity of peptide dimers by reducing the nonspecific DNA binding more effectively. For example, the synthetic dimerization domains described above possess a relatively flexible structure, which necessitates the design of a new enantiomeric domain with a more rigid structure and larger size to emphasize the steric constraints.

In addition to addressing the functional role of protein–protein interactions in the DNA–protein interactions, it is necessary to investigate structural aspects of DNA–protein recognition as well. It has been shown that the basic region of DNA-binding peptide does not form a specific structure in the absence of its specific DNA-binding sequence, but that it adopts an α-helical structure when bound to its target DNA sequence [86–91, 94, 105–107, 113, 116, 117]. This induced-fit mechanism has been thought to affect the specific DNA recognition of bZIP proteins [90, 91]. Because the basic region and leucine-zipper domains exist within the same polypeptide chain, the helix formation at the basic region would be promoted by the coiled-coil formation at the leucine zipper domain. In addition, the α-helical structure of the basic region would further stabilize the coiled-coil dimerization domain. Such a structural linkage between the DNA-binding region and the dimerization domain would affect the sequence-specific recognition of DNA. Thermodynamic and the kinetic comparison of the DNA-binding properties of these peptides with that of native bZIP proteins will clarify the effect(s) of structural linkages between the dimerization and DNA-binding domains in the sequence-selective recognition.

While the model studies described herein have thus far used DNA-binding domains derived from the native protein, the exact nature of how the DNA-binding motif of proteins fits into certain DNA sequences remains unanswered. The X-ray crystallographic and NMR studies provide detailed information on the stereospecific interactions between the amino acid residues and DNA. With this knowledge in mind, design and synthesis of a novel DNA-binding motif composed either on native or on non-natural amino acids will be undertaken. Such a study will be a crucial test of our understanding of molecular recognition, and the results will aid in expanding our knowledge of the sequence-specific DNA recognition mechanisms.

References

[1] N. Jones, *Cell* 61 (1990) 9.
[2] M. Ptashne, *A Genetic Switch*, Blackwell Scientific, Oxford, and Cell Press, Palo Alto, CA, 1986.
[3] C.J. Fry, P.J. Farnham, *J. Biol. Chem.* 274 (1999) 29583.
[4] W.P. Tansey, W. Herr, *Cell* 88 (1997) 729.
[5] R. Tijan, T. Maniatis, *Cell* 77 (1994) 5.
[6] T. Kodadek, *Chem. Biol.* 2 (1995) 267.
[7] C. Wilberger, *Annu. Rev. Biophys. Biomol. Struct.* 28 (1999) 29.
[8] T.K. Kim, T. Maniatis, *Mol. Cell* 1 (1997) 119.
[9] R. Grosschedl, *Curr. Opin. Cell Biol.* 7 (1995) 362.
[10] S.C. Harrison, A.K. Aggarwal, *Annu. Rev. Biochem.* 59 (1990) 933.
[11] C.O. Pabo, R.T. Sauer, *Annu. Rev. Biochem.* 61 (1992) 1053.
[12] A.K. Aggarwal, *Curr. Opin. Struct. Biol.* 5 (1995) 11.

[13] G. Patikoglou, S.K. Burley, *Annu. Rev. Biophys. Biomol. Struct.* 26 (1997) 289.

[14] B.W. Matthews, *Nature* 335 (1988) 294.

[15] R.G. Brennan, B.W. Matthews, *J. Biol. Chem.* 264 (1989) 1903.

[16] W.J. Gehring, M. Müller, M. Affolter, A. Percival-Smith, M. Billeter, Y.Q. Qian, G. Otting, K. Wüthrich, *Trends Genet.* 6 (1990) 323.

[17] S.C. Harrison, *Nature* 353 (1991), 715.

[18] R.G. Brennan, *Curr. Opin. Struc. Biol.* 1 (1991) 80.

[19] C.R. Kissinger, B. Liu, E. Martin-Blanco, T.B. Kornberg, C.O. Pabo, *Cell* 63 (1990) 579.

[20] C. Wolberger, A.K. Vershon, B. Liu, A.D. Johnson, C.O. Pabo, *Cell* 67 (1991) 517.

[21] W.J. Gehring, Y.Q. Qian, M. Billerter, K. Furukubo-Tokunaga, A.F. Schier, D. Resendez-Perez, M. Affolter, G. Otting, K. Wüthrich, *Cell* 78 (1994) 211.

[22] T. Li, M.R. Stark, A.D. Johnson, C. Wolberger, *Science* 270 (1995) 262.

[23] S. Tan, T.J. Richmond, *Nature* 391 (1998) 660.

[24] W.H. Landschultz, P.F. Johnson, S.L. McKnight, *Science* 240 (1988) 1759.

[25] C.R. Vinson, P.B. Sigler, S.L. McKnight, *Science* 246 (1989) 911.

[26] H.C. Hurst, *Protein Profile* 1 (1994) 123.

[27] A.G. Hinnebusch, *Proc. Natl. Acad. Sci. USA* 81 (1984) 6442.

[28] W.H. Landschultz, P.F. Johnson, E.Y. Adashi, B.J. Graves, S.L. McKnight, *Genes. Dev.* 2 (1988) 786.

[29] G.C. Prendergast, E.B. Ziff, *Nature* 341 (1989) 392.

[30] C. Murre, P.S. McCaw, D. Baltimore, *Cell* 56 (1989) 777.

[31] A.B. Lassar, J.N. Buskin, D. Lockson, R.L. Davis, S. Apone, S.D. Hauschka, H. Weintraub, *Cell* 58 (1989) 823.

[32] R.L. Davis, P.-F. Cheng, A.B. Lassar, H. Weintraub, *Cell* 60 (1990) 733.

[33] X.-H. Sun, D. Baltimore, *Cell* 64 (1991) 459.

[34] S.J. Anthony-Cahill, P.A. Benfield, R. Fairman, Z.R. Wasserman, S.L. Brenner, W.F. Stafford III, C. Altenbach, W. Hubbell, W.F. Degrado, *Science* 255 (1992) 979.

[35] T. Ellenberger, D. Fass, M. Arnaud, S.C. Harrison, *Genes Dev.* 8 (1994) 970.

[36] P.C.M. Ma, M.A. Rould, H. Weintraub, C.O. Pabo, *Cell* 77 (1994) 451.

[37] J.M. Berg, *Science* 232 (1986) 485.

[38] A. Klug, D. Rhodes, *Trends Biochem.* 12 (1987) 464.

[39] B.L. Vallee, J.E. Coleman, D.S. Auld, *Proc. Natl. Acad. Sci. USA* 88 (1991) 999.

[40] N.P. Pavletich, C.O. Pabo, *Science* 252 (1991), *252*, 809.

[41] R. Marmorstein, M. Carey, M. Ptashne, S.C. Harrison, *Nature* 356 (1992) 408.

[42] P.J. Kraulis, A.R.C. Raine, P.L. Gadhavi, E.D. Lane, *Nature* 356 (1992) 448.

[43] J.P. Baleja, R. Marmorstein, S.C. Harrison, G. Wagner, *Nature* 356 (1992) 450.

[44] N.P. Pavletich, C.O. Pabo, *Science* 261 (1993) 1701.

[45] S. Adhya, *Annu. Rev. Genet.* 23 (1989) 227–250.

[46] M. Brenowitz, N. Mandal, A. Pickar, E. Jamison, S. Adhya, *J. Biol. Chem.* 266 (1991) 1281.

[47] D. Beckett, D.S. Burz, G.K. Ackers, R.T. Sauer, *Biochemistry* 32 (1993) 9073.

[48] J. Chen, K.S. Matthews, *Biochemistry* 33 (1994) 8728.

[49] P.B. Dervan, *Curr. Opin. Chem. Biol.* 3 (1999) 688.

[50] S. White, J.W. Szewczyk, J.M. Turner, E.E. Baird, P.B. Dervan, *Nature* 391 (1998) 468.

[51] H.E. Moser, P.B. Dervan, *Science* 238 (1987) 645.

[52] J.R. Desjarlais, J.M. Berg, *Proc. Natl. Acad. Sci. USA* 90 (1993) 2256.

[53] Y. Choo, A. Klug, *Proc. Natl. Acad. Sci. USA* 91 (1994) 11168.

[54] E. Rebar, C.O. Pabo, *Science* 263 (1994) 671.

[55] A.C. Jamieson, S.-H. Kim, J.A. Wells, *Biochemistry* 33 (1994) 5689.

[56] H. Wu, W.-P. Yang, C.F. Barbas III, *Proc. Natl. Acad. Sci. USA* 92 (1995) 344.

[57] H.A. Greisman, C.O. Pabo, *Science* 275 (1997) 657.

[58] S.A. Wolfe, H.A. Greisman, E.L. Ramm, C.O. Pabo, *J. Mol. Biol.* 285 (1999) 1917.

[59] W.F. Anderson, D.H. Ohlendorf, Y. Takeda, B.W. Matthews, *Nature* 290 (1981) 754.

[60] D.H. Ohlendorf, W.F. Anderson, R.G. Fisher, Y. Takeda, B.W. Matthews, *Nature* 298 (1982) 718.

[61] D.B. Mckay, T.A. Steitz, *Nature* 290 (1981) 744.

[62] S.R. Jordan, C.O. Pabo, *Science* 242 (1988) 893.

[63] N.D. Clarke, L.J. Beamer, H.R. Goldberg, C. Berkower, C.O. Pabo, *Science* 254 (1991) 267.

[64] A. Laughon, M.P. Scott, *Nature* 310 (1984) 25.

[65] M.P. Scott, J.W. Tamkun, G.W. Hartzell, *Biochim. Biophys. Acta* 989 (1989) 25.

[66] J.C.W. Shepherd, W. McGinnis, A.E. Carrasco, E.M. De Robertis, W.J. Gehring, *Nature* 310 (1984) 70.

[67] R.T. Sauer, D.L. Smith, A.D. Johnson, *Genes Dev.* 2 (1988) 807.

[68] M. Affolter, A. Percival-Smith, M. Muller, W. Leupin, W.J. Gehring, *Proc. Natl. Acad. Sci. USA* 87 (1990) 4093.

[69] M.A. Garcia-Blanco, R.G. Clerc, P.A. Sharp, *Genes Dev.* 3 (1989) 739.

[70] M.F. Bruist, S.J. Horvath, L.E. Hood, T.A. Steitz, M.I. Simon, *Science* 235 (1987) 777.

[71] J.P. Sluka, S.J. Horvath, M.F. Bruist, M.I. Simon, P.B. Dervan, *Science* 238 (1987) 1129.

[72] D.P. Mack, B.I. Iverson, P.B. Dervan, *J. Am. Chem. Soc.* 110 (1988) 7572.

[73] D.P. Mack, P.B. Dervan, *J. Am. Chem. Soc.* 112 (1990) 4604.

[74] D.P. Mack, J. Shin, J.P. Sluka, J.H. Griffin, M.I. Simon, P.B. Dervan, *Biochemistry* 29 (1990) 6561.

[75] D.P. Mack, P.B. Dervan, *Biochemistry* 31 (1992) 9399.

[76] M.C. Mossing, R.T. Sauer, *Science* 250 (1990) 1712.

[77] H. Mihara, E.T. Kaiser, *Science* 242 (1988) 925.

[78] Z. Shang, V.E. Isaac, H. Li, L. Patel, K.M. Catron, T. Curran, G.T. Montelione, C. Abate, *Proc. Natl. Acad. Sci. USA* 91 (1994) 8373.

[79] T.E. Ellenberger, C.J. Brandl, K. Struhl, S.C. Harrison, *Cell* 71 (1992) 1223.

[80] P. König, T.J. Richmond, *J. Mol. Biol.* 233 (1993) 139.

[81] W. Keller, P. König, T.J. Richmond, *J. Mol. Biol.* 254 (1995) 657.

[82] J.W. Sellers, A.C. Vincent, K. Struhl, *Mol. Cell. Biol.* 10 (1990) 5077.

[83] S.J. Metallo, A. Schepartz, *Chem. Biol.* 1 (1994) 143.

[84] D.N. Paolella, C.R. Palmer, A. Schepartz, *Science* 264 (1994) 1130.

[85] E.K. O'Shea, J.D. Klemm, P.S. Kim, T. Alber, *Science* 254 (1991) 539.

[86] M.A. Weiss, T. Ellenberger, C.R. Wobble, J.P. Lee, S.C. Harrison, K. Struhl, *Nature* 347 (1990) 575.

[87] M.A. Weiss, *Biochemistry* 29 (1990) 8020.

[88] K.T. O'Neil, J.D. Shuman, C. Ampe, W.F. DeGrado, *Biochemistry* 30 (1991) 9030.

[89] V. Saudek, H.S. Pasley, T. Gibson, H. Gausepohl, R. Frank, A. Pastore, *Biochemistry* 30 (1991) 1310.

[90] N.P. Johnson, J. Lindstrom, W.A. Baase, P.H. von Hippel, *Proc. Natl. Acad. Sci. USA* 91 (1994) 4840.

[91] A.D. Frankel, P.S. Kim, *Cell* 65 (1991) 717.

[92] J.N. Glover, S.C. Harrison, *Nature* 373 (1995) 257, 261.

[93] C. Park, J.L. Campbell, W.A. Goddard III, *J. Am. Chem. Soc.* 118 (1996) 4235.

[94] T. Morii, J. Yamane, Y. Aizawa, K. Makino, Y. Sugiura, *J. Am. Chem. Soc.* 118 (1996) 10011.

[95] D. Stanojevic, G.L. Nerdine, *Nat. Struct. Biol.* 2 (1995) 450.

[96] S.J. Metallo, A. Schepartz, *Nature Struct. Biol.* 4 (1997) 115.

[97] P.H. Von Hippel, O.G. Berg, *Proc. Natl. Acad. Sci. USA* 83 (1986) 1608.

[98] N.Y. Sardesai, K. Zimmermann, J.K. Barton, *J. Am. Chem. Soc.* 116 (1994) 7502.

[99] N.Y. Sardesai, S.C. Lin, K. Zimmermann, J.K. Barton, *Bioconjugate Chem.* 6 (1995) 302.

[100] N.Y. Sardesai, J.K. Barton, *J. Biol. Inorg. Chem.* 2 (1997) 762.

[101] C.A. Hastings, J.K. Barton, *Biochemistry* 38 (1999) 10042.

[102] N.J. Zondle, A. Schepartz, *J. Am. Chem. Soc.* 121 (1999) 6938.

[103] R.V. Talanian, C.J. McKnight, P.S. Kim, *Science* 249 (1990) 769.

[104] R.V. Talaninan, C.J. McKnight, R. Rutkowski, P.S. Kim, *Biochemistry* 31 (1992) 6871.

[105] B. Cuenoud, A. Schepartz, *Science* 259 (1993) 510.

[106] M. Okagami, M. Ueno, K. Makino, M. Shimomura, I. Saito, T. Morii, Y. Sugiura, *Bioorg. Med. Chem.* 3 (1995) 777.

[107] M. Ueno, A. Murakami, K. Makino, T. Morii, *J. Am. Chem. Soc.* 115 (1993) 12575.

[108] C. Park, J.L. Campbell, W.A. Goddard III, *Proc. Natl. Acad. Sci. USA* 89 (1992) 9094.

[109] C. Park, J.L. Campbell, W.A. Goddard III, *Proc. Natl. Acad. Sci. USA* 90 (1993) 4892.

[110] C. Park, J.L. Campbell, W.A. Goddard III, *J. Am. Chem. Soc.* 117 (1995) 6287.

[111] B. Cuenoud, A. Schepartz, *Proc. Natl. Acad. Sci. USA* 90 (1993) 1154.

[112] C.R. Palmer, L.S. Sloan, J.C. Adrian, B. Cuenoud, D.N. Paolella, A. Schepartz, *J. Am. Chem. Soc.* 117 (1995) 8899.

[113] T. Morii, S. Morimoto, I. Saito. *J. Inorg. Biochem.* 43 (1991) 468.
[114] T. Morii, M. Shimomura, S. Morimoto, I. Saito, *Nucleic Acids Sym. Series* 25 (1991) 185.
[115] T. Morii, M. Shimomura, M. Morimoto I. Saito, *J. Am. Chem. Soc.* 115 (1993) 1150.
[116] T. Morii, Y. Saimei, M. Okagami, K. Makino, Y. Sugiura, *J. Am. Chem. Soc.* 119 (1997) 3649.
[117] M. Ueno, M. Sawada, K. Makino, T. Morii, *J. Am. Chem. Soc.* 116 (1994) 11137.
[118] M. Pellegrini, R.H. Ebright, *J. Am. Chem. Soc.* 118 (1996) 5831.
[119] E.T. Kaiser, *Trends Biochem. Sci.* 12 (1987) 305.
[120] M. Mutter, S. Vuilleumier, *Angew. Chem. Int. Ed. Engl.* 28 (1989) 535.
[121] W.F. DeGrado, Z.R. Wasseman, J.D. Lear, *Science* 243 (1989) 622.
[122] J.S. Richardson, D.C. Richardson, *Trends Biochem. Sci.* 14 (1989) 304.
[123] M. Lieberman, T. Sasaki, *J. Am. Chem. Soc.* 113 (1991) 1470.
[124] M.R. Ghadiri, C. Soares, C. Choi, *J. Am. Chem. Soc.* 114 (1992) 825.
[125] A. Schepartz, J.D. McDevitt, *J. Am. Chem. Soc.* 111 (1989) 5976.
[126] T. Morii, A. Murakami, K. Makino, S. Morimoto, I. Saito, *Tetrahedron Lett.* 35 (1994) 1219.
[127] P.F. Johnson, *Mol. Cell. Biol.* 13 (1993) 6919.
[128] J. Kim, D. Tzamarias, T. Ellenberger, K. Struhl, S.C. Harrison, *Proc. Natl. Acad. Sci. USA* 90 (1993) 4513.
[129] T. Sera, P.G. Schultz, *Proc. Natl. Acad. Sci. USA* 93 (1996) 2920.
[130] M. Brenowitz, N. Mandal, A. Pickar, E. Jamison, S. Adhya, *J. Biol. Chem.* 266 (1991) 1281.
[131] D. Beckett, D.S. Burz, G.K. Ackers, R.T. Sauer, *Biochemistry* 32 (1993) 9073.
[132] J. Chen, K.S. Matthews, *Biochemistry* 33 (1994) 8728.
[133] M.R. Eftink, M.L. Andy, K. Bystrom, H.D. Perlmutter, D.S. Kristol, *J. Am. Chem. Soc.* 111 (1989) 6765.
[134] Y. Inoue, T. Hakushi, Y. Liu, L.-H. Tong, B.-J. Shen, D.-S. Jin, *J. Am. Chem. Soc.* 115 (1993) 475.
[135] K.A. Conners, *Chemical Reviews* 97 (1997) 1325.
[136] Y. Inoue, *Chemical Reviews* 98 (1998) 1875.
[137] T.K. Kerppola, T. Curran, *Curr. Opin. Struc. Biol.* 1 (1991) 71.
[138] J.L. Pomerantz, P.A. Sharp, C.O. Pabo, *Science* 267 (1995) 93.
[139] J.L. Pomerantz, C.O. Pabo, P.A. Sharp, *Proc. Natl. Acad. Sci. USA* 92 (1995) 9752.
[140] J.S. Kim, J. Kim, K.L. Cepek, P.A. Sharp, C.O. Pabo, *Proc. Natl. Acad. Sci. USA* 94 (1997) 3616.
[141] J.L. Pomerantz, S.A. Wolfe, C.O. Pabo, *Biochemistry* 37 (1998) 9665.
[142] R. Marmorstein, M. Carey, M. Ptashne, S.C. Harrison, *Nature* 356 (1992) 408.
[143] B.S. Wang, C.O. Pabo, *Proc. Natl. Acad. Sci. USA* 96 (1999) 9568–9573
[144] Y. Aizawa, Y. Sugiura, M. Ueno, Y. Mori, K. Imoto, K. Makino, T. Morii, *Biochemistry* 38 (1999) 4008.
[145] Y. Aizawa, Y. Sugiura, T. Morii, *Biochemistry* 38 (1999) 1626.
[146] J.S. Kim, C.O. Pabo, *Proc. Natl. Acad. Sci. USA* 95 (1998) 2812.
[147] R.R. Beerli, D.J. Segal, B. Dreier, C.F. Barbas III, *Proc. Natl. Acad. Sci. USA* 95 (1998) 14628.
[148] G.K. Ackers, A.D. Johnson, M.A. Shea, *Proc. Natl. Acad. Sci. USA* 79 (1982) 1129.
[149] T.J. Smith, *Molview* Department of Chemistry, Purdue University, 1994.

Advances in DNA Sequence-specific Agents 04 (2002) 139–155

Equilibrium and kinetic quantitative DNase I footprinting

Gauri M. Dhavan, A. K. M. M. Mollah,
Michael Brenowitz*

*Department of Biochemistry, Albert Einstein College of Medicine,
1300 Morris Park Avenue, Bronx, NY 10461*

1. Introduction

Quantitative DNase I footprinting is a powerful tool for probing the equilibrium binding and kinetic rate constants that describe protein binding to one or more sites on DNA. The ability of this technique to yield thermodynamically valid individual-site binding isotherms was first demonstrated for the interaction of bacteriophage lambda c*I* repressor with its three specific sites within the right operator of the phage genome [1–3]. These and subsequent studies provide the most comprehensive illustration of individual-site binding analysis applied to the resolution of microscopic binding constants that describe the intrinsic affinity of proteins for specific sites and cooperative interactions among them. General discussions of individual-site binding theory [4, 5] and detailed experimental protocols for this technique [6–8] have been published. Quantitative DNase I footprinting protocols have also been developed for conducting quench-flow kinetics studies of protein–DNA interactions [9, 10]. These approaches have been applied to a variety of systems and experimental questions [11–14].

This chapter uses two DNA-binding proteins, one prokaryotic and one eukaryotic, to illustrate the application of quantitative DNase I footprinting to cellular regulatory systems. The first protein is the TATA box binding protein (TBP) from *Saccharomyces cerevisiae*. TBP is required for the initiation of transcription by all three eukaryotic RNA polymerases [15]. Its binding to specific promoter sequences called "TATA boxes" is a crucial step in the initiation of transcription of genes transcribed by RNA polymerase II [16]. Co-crystal structures of TBP bound to DNA containing TATA box sequences [17–21] reveals that the DNA is significantly distorted from normal B-DNA within the TBP-TATA complexes.

---

*Corresponding author
E-mail address: brenowit@aecom.yu.edu
Advances in DNA Sequence-specific Agents, Volume 4 G. B. Jones (Editor)

The second protein is the Integration Host Factor (IHF) from *Escherichia coli*. IHF binds a number of bacterial and lambda phage DNA sequences and is an accessory protein in transcription, DNA replication, phage λ packaging and site-specific recombination [22–25]. IHF binding to the phage λ genome is required for the formation of a nucleo-protein complex that is a necessary step in phage integration into the host genome [26]. The phage λ DNA contains three IHF binding sites, H', H1, and H2, and, of the three, IHF exhibits the highest affinity for the H' site while protecting around 50 bp of DNA from DNase I attack [27,28]. The protein makes several direct contacts between amino acid side chains and DNA bases or phosphates on the nucleic acid backbone [27, 29].

Although both IHF and TBP bind in the minor groove and impart a significant bend in the bound DNA, they do so in very different fashions [17–21,29]. Monomeric TBP binds to an 8 bp sequence, unwinds the bound minor groove and bends the DNA away from itself. In contrast, IHF is a heterodimeric protein that interacts with nearly 50 bp of DNA making extensive contacts in the minor groove and wrapping the DNA around it into a hairpin bend.

The quantitative DNase I footprinting methods being used to probe the thermodynamics and kinetics of these similar, yet distinct, protein–DNA interactions are presented in this chapter.

2. Equilibrium DNase I footprinting

The intrinsic free energy of protein binding to its DNA recognition sequence can be obtained by measuring the equilibrium-binding constant, K_{eq}, for protein–DNA complex formation. The relationship between the free energy of an association event, $\Delta G°$, and K_{eq} is expressed by the equation $\Delta G° = -RT \ln K_{eq}$. Achieving an understanding of the energetics of protein–DNA association reactions requires the accurate determination of K_{eq} over a range of solution conditions such as pH and ionic strength and external variables like temperature. In doing so, a thorough thermodynamic understanding of the enthalpic and entropic contributions to the free energy of binding can be assessed for any DNA–protein system. The effectiveness of quantitative footprinting in measuring individual site equilibrium binding constants and binding free energies of each protein–DNA contact in a multi-protein–DNA complex has been demonstrated [1–3].

Additionally, the quantitative footprinting technique offers a straightforward approach to compare the binding of the same protein to different DNA sequences, the binding of multiple proteins in a cooperative fashion to adjacent sites on DNA, or a means to study the effect of mutations on the energetics of protein binding to its recognition sequence. The use of radiolabeled DNA in this assay allows very high affinity interactions to be accurately investigated.

Quantitative DNase I footprint titrations are conducted in solution by equilibrating a series of protein concentrations with a constant concentration of radiolabeled DNA. After the appropriate concentrations of protein and DNA are mixed in each reaction tube, the mixture is incubated at the desired temperature and solution conditions for sufficient time to allow the protein and DNA to reach equilibrium. "Footprinting" of the DNA by the addition of DNase I is the final step in the assay. The competition between the binding protein and the nuclease allows the determination of the extent of DNA protection at each protein concentration, providing a direct measure of binding equilibria at the exact point of DNA contact.

Protein   Proteins purified to homogeneity of known specific DNA-binding activity are a prerequisite for quantitative studies. Careful characterization of the properties of the protein (i.e. possible self-association) is essential for the accurate analysis and interpretation of thermodynamic and kinetic footprinting data.

DNA   A linear DNA restriction fragment derived from a plasmid that is labeled at either one 3′ or 5′ end with ^{32}P is a standard substrate for DNase I footprinting. The desired DNA can also be amplified by PCR using one 5′ labeled primer. PCR offers the advantage of easily changing the DNA fragment length by using different primers complementary to the plasmid sequence outside the cloned DNA and is not dependent on the presence of available restriction endonuclease sites along the plasmid. It should also be noted that "indirect labelling" methods such as primer extension can be used in footprinting assays. Indirect labeling or ligation of ^{32}P-labeled DNA allows covalently closed DNA substrates to be used in quantitative footprinting assays.

Preferably, a protein binding site(s) is located 30–90 bp from the ^{32}P-labeled end of the DNA although shorter and longer spacing can be successfully analyzed by using the appropriate electrophoretic conditions. An important consideration is the use of sufficiently long DNA substrates to minimize potential problems in DNA binding due to proximity of the protein binding site to end of the DNA molecule.

DNA labeling protocols

3′ end labeling of restriction fragments   The plasmid containing the binding site, as in the case with the TBP binding site in our studies, is digested with restriction endonuclease in order to generate a 3′ recessed end at the desired distance. The latter end is labeled with the appropriate ^{32}P-dNTPs using the large fragment of DNA polymerase I (Klenow fragment). Following the labeling step, a second restriction digest is applied that cleaves downstream of the protein-binding site in order to generate a restriction fragment labeled at a single end [8].

5′ end labeling of PCR amplified duplex   The particular region of interest can be amplified by PCR before radiolabeling. For the studies of IHF the H′ site was cloned into the plasmid puc19 (New England BioLabs) and amplified by PCR. One of the PCR primers is phosphorylated at the 5′ end. This procedure allows specific radiolabeling of the other end in the final PCR product with T4 Polynucleotide Kinase. This approach is a modification of the PCR amplification technique discussed in reference [30].

The equilibrium DNase I "footprinting" protocol

The appropriate concentration range probed to obtain the equilibrium binding constant must extend from zero to 100% saturation and can be determined by conducting a simulation of the appropriate binding isotherm if the apparent dissociation constant is known. DNase I footprinting can then be conducted to assay for the extent of binding at each concentration.

In our experiments the protein was serially diluted into microfuge tubes containing the desired buffer before adding the ^{32}P-DNA (at least 20,000 dpm per reaction). The final volume in all reaction tubes was 100 μL. The samples were thoroughly mixed and allowed to equilibrate in a temperature-regulated bath. An adequate equilibration time is important for this experiment as the associating molecules must be allowed to reach equilibrium. For most protein–DNA complexes an incubation period of 30 min to one hour is sufficient but, for molecules that have lower affinities for their binding site, longer equilibration periods may have to be applied.

Another important consideration in this experiment is that the DNA cut does not exceed "single-hit conditions" for reasons described in reference [2]. Also, the volume of DNase I stock added should be as small as possible so that the protein concentration is not appreciably reduced and the DNase I exposure time must be short relative to the equilibration time of the complex. In our experiments we added 5μl of DNase I, mixed the solution gently, and quenched the reaction after 2 min by the addition of 20 μl 50 mM EDTA following by vigorous mixing using a vortexer. Following the quench step, the sample was ethanol precipitated by conventional procedures. In recovering the sample by ethanol precipitation, care was taken to retrieve as much of the sample as possible, especially the smaller molecular weight DNA, so that the intensity of each band in a lane was an accurate representation of total cleaved DNA. For this reason we added excess tRNA or linear polyacrylamide during this step to increase nucleic acid precipitation efficiency.

Examples of equilibrium DNase I footprint titrations

Figure 1A shows a digital autoradiogram of a DNase I footprint titration experiment for the binding of TBP to DNA bearing the "TATA box" sequence of the adenovirus major late promoter, TATAAAAG. The increase in the protection of the DNA from DNase I cleavage at the TATA box is clearly evident in the autoradiogram. Quantitation of the protection is achieved in our experiments using the ImageQuant™ (Molecular Dynamics™) following published protocols [9]. A key step in the quantitation of protein binding is the standardization of extent protection observed in the binding-site relative to regions not protected by protein in order to compensate for differences in the total counts within each lane. This is accomplished by dividing the total counts measured for the protected binding site by the counts measured for a region outside the area contacted by protein.

Since the concentration of ^{32}P-labeled DNA used in footprinting experiments ($\leqslant 100$ pM) is typically smaller than the K_d, the approximation [protein]$_{total} \approx$ [protein]$_{free}$ can be used in the analysis of the titrations. Fractional saturation ($\bar{Y}$) of the protein binding-site is determined from the fractional protection (p_i) of bands visualized by nonlinear least squares fitting of the data against

$$p_i = p_{i,\text{lower}} + (p_{i,\text{upper}} - p_{i,\text{lower}}) \times \bar{Y} \tag{1}$$

where

$$\bar{Y} = K[P]/(1 + K[P]) \tag{2}$$

where $p_{i,\text{lower}}$ and $p_{i,\text{upper}}$ are the limits of the transition curves, respectively, K is the equilibrium association constant and $[P]$ is the concentration of protein. Each data set is analyzed and scaled to $\bar{Y}$ using the best-fit transition endpoints (eq. (1)) to yield the binding isotherm (Fig. 1B). When multiple data sets are globally analyzed, each data set was weighted by the inverse of the square root of the variance of its individual fit.

As can be seen in Figure 1B, the TBP-TATA interaction is well described by eq. (2), consistent with the binding of a single molecule of TBP to a single binding site on the DNA [31, 32]. The TBP concentrations used in this titration increased over several orders of magnitude, important for defining the upper and lower limits (eq. 1) of the titration transition. Extensive discussions of the issues that must be considered in analyzing quantitative nuclease protection experiments have been recently published [33, 34].

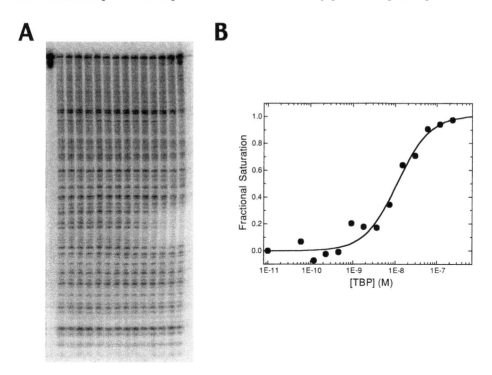

Fig. 1. (A) Digital autoradiogram of an equilibrium DNase I "footprint" titration for TBP binding to a single specific site on a 282 bp restriction fragment. The experiment was conducted in 25 mM Bis-tris, 5mM MgCl$_2$, 1 mM CaCl$_2$, 2mM DTT, 100mM KCl, 0.01% Brij, 1 μg/ml poly dG/dC at pH 6.5, and 25°C with an equilibration time of 45 min. The concentration of TBP monomer increases from 0–900 nM (left to right). The concentration of the ^{32}P-labeled DNA is ≤ 10 pM. (B) Binding isotherm derived from the experiment shown in A. The free energy ($\Delta G°$) of this interaction obtained by a least squares fit of the data to eqs. (1) and (2) is -11.0 ± 0.2 kcal/mol, which corresponds to $K_{eq} = 1.2 \times 10^8$.

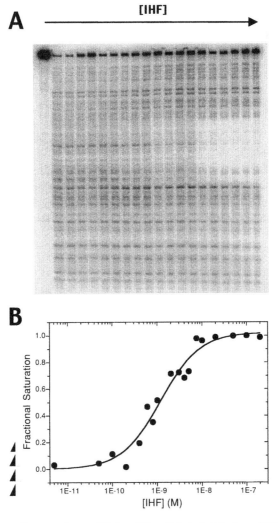

Fig. 2. (A) Digital autoradiogram of an equilibrium DNase I footprint titration of IHF binding at the H′ site. The protein–DNA incubation period was 30 min at 25°C in a buffer containing 50mM Tris-HCl, 70mM KCl, 7mM MgCl$_2$, and 3mM CaCl$_2$ at pH 7.0. The concentration of IHF ranges from 0 to 200 nM. The first lane contains uncut control DNA. The DNA concentration in each reaction was ~ 100 pM. (B) The least squares fit of the quantified change in footprint intensity of the experiment shown in A to eqs. (1) and (2). The free energy ($\Delta G°$) of this interaction is –12.1 ± 0.2 kcal/mol, which corresponds to $K_{eq} = 7.5 \times 10^8$.

The second example of a DNase I footprint titration experiment is a titration of IHF with duplex DNA containing the H′ binding sequence from the phage λ genome (Fig. 2A). As with the TBP titration, the IHF concentration is increased over several orders of magnitude, sufficient to clearly define the upper and lower limits to the transition. Quantitation of the extent of protection of the H′ site and fitting to eqs. (1) and (2) yields the binding isotherm shown in Figure 1B. The low DNA concentration allows the approximation $[IHF]_{total} \approx [IHF]_{free}$ to be made in this experiment. The isotherm is well described by eq. (2), consistent with the binding of a single stable heterodimer to the H' site on the DNA. The value of K_{eq} obtained from this experiment is in agreement with the affinity for the same sequence reported by the nitrocellulose filter-binding assay [35]. In contrast, the values reported by DNase I footprinting and filter-binding are an order of a magnitude less than that determined in gel mobility-shift experiments [36].

3. Protein–DNA association kinetics

Kinetics studies of protein–DNA interaction provide insight into the mechanism of sequence-specific recognition by proteins. Rates of association can vary widely among different protein–DNA complexes ranging from significantly slower than limited by diffusion [14, 31] to those in which "facilitated diffusion" along the DNA [37] result in rates that exceed diffusion by orders of magnitude. Similarly, dissociation reactions vary in rate by orders of magnitude. The measurement of these rates at specific sites on the DNA can provide an insight into the pathways of DNA–protein complex formation, as demonstrated in this section for the formation of the TBP-TATA and IHF-DNA complexes.

Quench-flow DNase I footprinting can monitor the appearance and disappearance of specific sites of protection on a millisecond timescale [9]. This ability to monitor the rates of protection at individual sites potentially allows the resolution of heterogeneity in association rates among multiple DNA contacts or within multi-protein–DNA complexes. In simple bimolecular complexes, such as those presented here, quench-flow DNase I footprinting yields rates for the initial binding event in complex formation. Since both IHF and TBP bend the DNA to which they bind, complex formation may involve a stepwise binding and bending process or be a concerted event in which binding and bending of the DNA occur virtually simultaneously. Quench-flow footprinting, in combination with kinetics assays that monitor global changes in DNA conformation, such as bending, can help define mechanisms of protein–DNA complex formation [38].

On a practical level, the lower protein concentrations that can be used to study the kinetics of high affinity protein–DNA interactions provide a slower time scale for concentration dependence kinetics, as is the case with bimolecular association events. Additionally, experiments can be designed to approximate "pseudo-first order" conditions where the ^{32}P-DNA substrate is present at a negligible concentration relative to the protein allowing the concentration of the other to determine the rate of association with single exponential saturation behavior. Further details regarding the reduction and analysis of kinetic data can be found in reference [9]. The quantitation of protected regions on a denaturing gel is conducted in a similar manner to that described above for the equilibrium binding titration assay.

The quench-flow apparatus provides the necessary automation in order to performed time-resolved footprinting of reactions with half-lives under a few minutes [39]. In the latter experiments the protein and DNA are mixed and allowed to bind for variable periods of time ranging from milliseconds to seconds before cutting with DNase I. Kinetic processes slower than this can be simply studied by mixing protein and DNA components manually (without the aid of a quench-flow device) and incubated for a period of time in the order of seconds before assaying for binding with a short DNase I reaction step.

Quench-flow DNase I footprinting

Experiments were conducted with a quench-flow apparatus allow mixing times under 100 msec. An equilibrated water bath surrounding the sample tubing, mixing chamber, and buffer syringes maintains the reactant and buffer solutions at the desired temperature. Since high concentrations of DNase I are generally used in kinetic experiments, contamination of samples is a common problem. Care should be taken to rinse all sample tubing with a quench solution (50 mM EDTA at pH 8.0) followed by de-ionized in between experiments.

A quench-flow system set up to perform time-resolved DNase I footprinting generally contains three reservoir syringes above the sample loops that are "primed" or filled with excess buffer. One of the three syringes is filled with the DNase I stock solution and is connected to a tube exiting the mixing chamber where the protein–DNA components are incubated during a kinetic experiment. A diagram of the quench-flow apparatus can be found in reference [9].

The DNA solution is fed into the apparatus via a side port (attached to a syringe) from where it can be introduced into a sample loop. The protein solution, which is initially at twice the final concentration desired upon mixing with the DNA, is also fed into a side port and introduced into a separate sample loop. The sample loops feed into a mixing chamber and, immediately before performing a kinetic time point, the flow control values of the quench-flow device are all turned to their "mixing" positions.

Mixing of the pre-loaded samples is achieved by pushing a predetermined volume of excess buffer into the sample loops, thus forcing the DNA and protein solutions into the mixing chamber. After allowing a predetermined pause during which the protein and DNA are allowed to bind, the mixture is pushed into the exit loop where it mixes with the DNase I solution and exits into a microfuge tube containing quench solution (50mM EDTA at pH 8.0) and a phenol–chloroform mixture. A computer-controlled motor gener- ates the "push-pause-push" cycling of the stepping motor used to conduct the experiment. The first "push" is conducted at the minimum speed required for adequate mixing under turbulent flow. The "pause" which enables the reaction to proceed for a defined mixing time period is the only variable changed during an experiment. The speed of the second "push", which mixes the protein–DNA reaction mixture with the DNase I, can be manip- ulated to change to the DNase I cutting time. Our standard protocol uses a 21 msec DNase I exposure.

After conducting one quench-flow cycle, the sample, reaction, and exit loops are cleaned by sequentially washing with quench solution (50mM EDTA at pH 8.0), doubly de-ionized water and methanol. The latter procedure prepares the apparatus for the collec- tion of a second kinetic time point (at a different mixing time).

The time-resolved DNase I footprinting experiment used to measure TBP-DNA asso- ciation kinetics is shown in Figure 3B. In this experiment the concentration of DNA was under 10 pM allowing the reaction kinetics to proceed under a psuedo-first order condi- tion where the concentration of the protein (which is in vast excess over DNA) determines the rate of the association process. Under these reactant concentrations the kinetic data can be fit to a single exponential as described by

$$\bar{Y} = 1 - e^{-k_{obs}t} \qquad (3)$$

where k_{obs} is the pseudo-second order rate constant and t is time. Data obtained from the gel was used to obtain a progress curve with a $k_{obs} = 0.11 \pm 0.2$ sec^{-1} (Fig. 3B). For a bimolecular association reaction $k_{obs} = k_a[TBP] + k_d$, where k_a is the second-order rate constant while k_d is the dissociation rate constant [40]. Direct global fitting of families of progress curves conducted at a series of protein concentrations can be used to determine values of k_a and k_d. (Alternatively, k_d can be independently measured in relaxation experiments and these experiments included in a global analysis.) When multiple data sets were globally analyzed, each data set is weighted by the inverse of the square root of the variance of its individual fit. The slower than diffusion limited association rate observed for TBP is proposed to be due to the extremely unfavorable quasi-equilibrium for the formation of a productive encounter between DNA and TBP [38].

The association of IHF with its binding sites is most likely the first event in several cellular processes in which DNA bending is a key phenomenon. IHF's role as an accessory protein recruited for the purpose of imposing a severe alteration in DNA structure during recombination, transcription, and DNA replication is well documented [22–25]. For this reason, the rate of IHF association will determine the overall rate of biological processes in which it figures as a necessary participant.

In Figure 4 we demonstrate the association kinetics of IHF binding to one of its sequences from the phage λ genome (the H' binding site). Two time-resolved DNase I footprinting experiments were conducted using IHF concentrations of 10 and 20nM. The DNA concentration was kept low (~ 100pM) so that the fractional saturation of IHF binding could be fit to a psuedo-first order approximation, as discussed earlier for the association of TBP with its binding site. The pseudo-second order rate constant of IHF binding, k_{obs}, measured is 3.5 sec^{-1} or higher defining it on the order of, or beyond, diffusion. The k_{obs} measured for at IHF concentration of 10 nM defines a lower limit for DNA–protein association since the rate obtained at higher IHF concentrations (as demonstrated for a protein concentration of 20nM) at this temperature and buffer conditions is faster than the resolution time of this experiment. The latter discovery suggests that IHF-DNA sequence-specific complex formation involves a facilitated diffusion step in which the protein searches for its specific binding sequence. A similar model has been proposed for the Lac repressor suggesting that the protein slides along the DNA helix in a one dimensional diffusion step before encountering its sequence-specific binding site [40]. Though it is not clear if such is the case for the binding between IHF and DNA, the faster than diffusion association rate can only be achieved via some manner of facilitated diffusion. In order to define the exact mechanism of sequence-specific site recognition by IHF, a more detailed study of the dependence of association on the variables of ionic strength, temperature, and DNA length will be required.

4. Conclusion

Footprinting techniques allow the binding of proteins at their exact points of contact on the DNA to be monitored. When used in combination with other assays, quantitative

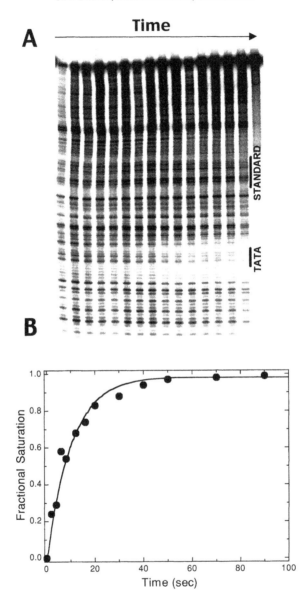

Fig. 3. (A) Digital image of an electrophoretogram of a quench-flow "footprinting" experiment for TBP binding to a single specific site, TATA on the 282 bp cloned adenovirus major late promoter. The experiment was conducted in the buffer described in the legend to Figure 1 at pH 7.0 and 30°C. The concentration of TBP was 300 nM (monomer) and the ^{32}P-labeled DNA <1pM. (B) Individual-site progress curve of the experiment shown in A. The value of ka determined for this experiment and a set of similar experiments conducted as a function of TBP concentration was $3.5 \pm 0.8 \times 10^5$ M^{-1}sec^{-1}.

footprinting offers the ability to resolve intermediates in the kinetic pathways of sequence-specific protein–DNA complex formation. In a similar manner, quantitative footprinting offers a direct approach for resolving the energetic contributions of individual proteins that bind DNA in a cooperative fashion. While the number of DNA base pairs protected varies for different protein/ligand–DNA complexes, this does not affect the accuracy of this technique, as shown here for TBP and IHF, since the extent of protection as a func-

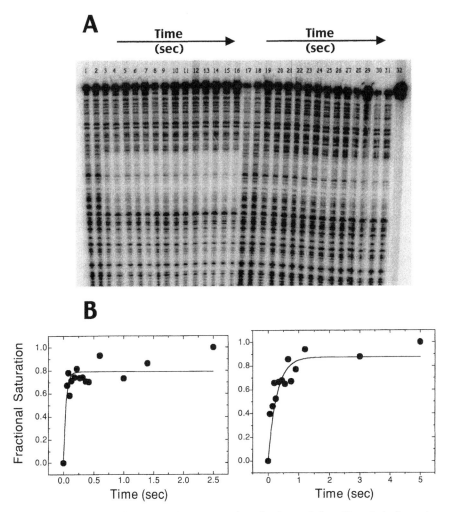

Fig. 4. (A) Digital image of an electrophoretogram of a pair of quench-flow "footprinting" experiments for IHF with the H′ site. The first sixteen lanes represent a kinetic time course at an IHF concentration of 20 nM and the remainder for a concentration of 10 nM. The time points range from 0 to 2.5 sec for the higher IHF concentration and 0 to 5 sec for an IHF concentration of 10 nM. The last lane contains uncut DNA. The plots in (B) represent the least squares fits to eq. (3) of the fractional saturation (Yb) at the H′ site as a function of time. The lower limit for k_{obs} obtained by the fit of the time course of binding at an IHF concentration of 10 nM is 3.5 sec^{-1} and defines a lower limit for the psuedo second order rate.

tion of protein concentration or time is a relative measure. Quantitative DNase I footprinting is a versatile technique for the analysis of both simple and complex assemblies of protein on DNA.

References

[1] M. Brenowitz, D.F. Senear, M.A. Shea, G.K. Ackers, Footprint titrations yield valid thermodynamic isotherms. *Proc. Nat. Acad. Sci.* 83, (1986) 8462–8466.

[2] M. Brenowitz, D.F. Senear, M.A. Shea, G.K. Ackers, Quantitative DNase footprint titration: A method for studying protein–DNA interactions. *Meth. Enzymol.* 130 (1986) 132–181.

[3] D.F. Senear, M. Brenowitz, G.K. Ackers, Energetics of cooperative protein–DNA interactions: Comparison between quantitative DNase footprint titration and filterbinding. *Biochemistry* 25 (1986) 7344–7354.

[4] G.K. Ackers, M.A. Shea, F.R. Smith, Free energy coupling within macromolecules. The chemical work of ligand binding at the individual sites in co-operative systems. *J. Mol. Biol.* 170 (1983) 223–242.

[5] E. Di Cera, *Thermodynamic Theory of Site-Specific Binding Processes in Biological Macromolecules*, Cambridge University Press, Cambridge, 1983.

[6] M. Brenowitz, D.F. Senear, DNase I footprint analysis of protein–DNA binding. In: F.M. Ausubel, R. Brent, R.E. Kingston, D.D. Moore, J.G. Seidman, J.A. Smith, K. Struhl, eds, *Current Protocols in Molecular Biology* (Unit 12.4). John Wiley & Sons, New York, 1989.

[7] Brenowitz, M., Senear, D.F., Jamison, L., Dalma-Weiszhausz, D.D. Quantitative DNase I footprinting. In: A. Revzin, ed., *Footprinting Techniques for Studying Nucleic Acid-Protein Complexes* (a volume of *Separation, Detection, and Characterization of Biological Macromolecules*) (pp. 1–43). Academic Press, New York, 1993.

[8] Mollah, A.K.M.M., Brenowitz, M. Quantitative DNase I kinetics footprinting. Protein-DNA interactions — a practical approach. In: A. Travers, M. Buckle, eds, IRL Press at Oxford University Press, Oxford, 2000, pp. 281–290.

[9] M. Hsieh, M. Brenowitz, Quantitative kinetics footprinting of protein–DNA association reactions. *Meth. Enzymol.* 274 (1996) 478–492.

[10] M. Hsieh, M. Brenowitz, Comparison of the DNA association kinetics of the Lac repressor tetramer, its dimeric mutant LacI^{adi} and the native dimeric gal repressor. *J. Biol. Chem.* 272 (1997) 22092.

[11] V. Petri, M. Brenowitz, Quantitative nucleic acids footprinting — thermodynamic and kinetic approaches. *Current Opinion in Biotechnology* 8 (1997) 36–44.

[12] T.R. Pray, D.S. Bruz, G.K. Ackers, Cooperative non-specific DNA binding by octamerizing λ cI repressors: A site-specific thermodynamic analysis. *J. Mol. Biol.* 282 (1998) 947–958.

[13] E.D. Streaker, D. Beckett, Coupling of site-specific DNA binding to protein dimerization in assembly of the biotin repressor-biotin operator complex. *Biochemistry* 37 (1998) 3210–3219.

[14] M.L. Craig, O.V. Tsodikov, K.L. McQuade, P.E. Schlax Jr, M.W. Capp, R.M. Saecker, T.M. Record Jr, DNA footprints of the two kinetically significant intermediates in formation of an RNA polymerase-promoter open complex: Evidence that interactions with start site and downstream DNA induce sequential conformational changes in polymerase and DNA. *J. Mol. Biol.* 283 (1998) 741–756.

[15] S. McKnight, Transcription revisited: A commentary on the 1995 Cold Spring Harbor Laboratory meeting, "Mechanisms of Eukaryotic Transcription". *Genes and Development* 10 (1996) 367–381.

[16] S.K. Burley, R.G. Roeder, Biochemistry and structural biology of transcription factor IID (TFIID). *Annul. Rev. Biochem.* 65 (1996) 760–799.

[17] Y. Kim, J.H. Geiger, S. Hahn, P.B. Sigler, Crystal structure of a yeast TBP/TATA-box complex. *Nature* 365 (1993b) 512–520.

[18] J.L. Kim, S.K. Burley, 1.9 A resolution refined structure of TBP recognizing the minor groove of TATAAAAG. *Nature Struct. Biol.* 1 (1994a) 638–653.

[19] Z.S. Juo, T.K. Chiu, P.M. Leibermann, I. Baikalov, A.J. Berk, R.E. Dickerson, How proteins recognize the TATA box. *J. Mol. Biol.* 261 (1996) 239–254.

[20] D.B. Nikolov, H. Chen, E.D. Halay, A. Hoffman, R.G. Roeder, S.K. Burley, Crystal structure of a human TATA box-binding protein/TATA element complex. *Proc. Natl. Acad. Sci. USA* 93 (1996) 4862–4867.

[21] G.A Patikogluou, J.L. Kim, L. Sun, S.H. Yang, T. Kodadek, S.K. Burley, TATA element recognition by the TATA box-binding protein has been conserved throughout evolution. *Genes and Development 15* 13 (1999) 3217–3230.

[22] J. Gardner, H.A. Nash, Role of *Escherichia coli* IHF protein in lambda site-specific recombination. *J. Mol. Biol.* 191 (1986) 181–189.

[23] W. Zin, M. Feiss, Function of IHF in λ DNA packaging. I. Identification of the strong binding site of integration host factor and the locus for intrinsic bending in *cosB. J. Mol. Biol.* 230 (1993) 492–504.

[24] D.S. Hwang, A. Kornberg, Opening of the replication origin of *Escherichia coli* by DnaA protein with protein HU or IHF. *J. Biol. Chem.* 267 (1992) 23083–23086.

[25] J.W. Winkelman, G.W. Hatfield, Characterization of the integration host factor binding site in the ilvPG1 promoter region of the ilvGMEDA operon of *Escherichia coli. J. Biol. Chem.* 265 (1990) 10055–10060.

[26] N. Craig, H.A. Nash, The mechanism of phage 1 site-specific recombination: site-specific breakage of DNA by Int Topoisomerase. *Cell* 35 (1983) 795–803.

[27] S. Yang, H.A. Nash, The interaction of *E. coli* IHF protein with its specific binding sites. *Cell* 57 (1989) 869–880.

[28] S. Yang, H.A. Nash, Comparison of protein binding to DNA *in vivo* and *in vitro*: Defining an effective intracellular target. *EMBO J.* 14 (1995) 6292–6300.

[29] P.A. Rice, S. Yang, K. Mizuuchi, H.A. Nash, Crystal structure of an IHF-DNA complex: a protein-induced DNA U-turn. *Cell* 87 (1996) 1295–1306.

[30] C. Bailly, D. Payet, A.A. Travers, M.J. Waring, PCR-based development of DNA substrates containing modified bases: An efficient system for investigating the role of the exocyclic groups in chemical and structural recognition by minor groove binding drugs and proteins. *Proc. Natl. Acad. Sci. USA* 93 (1996) 13623–13628.

[31] V. Petri, M. Hsieh, M. Brenowitz, Thermodynamic and kinetic characterization of the binding of the TATA binding protein to the adenovirus E4 promoter. *Biochemistry* 34 (1995) 9977–9984.

[32] V. Petri, M. Hsieh, E. Jamison, M. Brenowitz, DNA Sequence-specific recognition by the "TATA" binding protein: Promoter dependent differences in the thermodynamics and kinetics, *Biochemistry* 37 (1998) 15842–15849.

[33] D.F. Senear, D.W. Bolen, Simultaneous analysis for testing of models and parameter estimation. *Meth. Enzymol.* 210 (1992) 463–481.

[34] K.S. Koblan, D.L. Bain, D. Beckett, M.A. Shea, G.K. Ackers, Analysis of site-specific interaction parameters in protein–DNA complexes. *Meth. Enzymol.* 210 (1992) 405–425.

[35] H. Kurumizaka, F. Kanke, U. Matsumoto, H. Shindo, Specific and nonspecific interactions of integration host factor with oligo DNAs and revealed by circular dichroism specitroscopy and filter binding assay. *Archives of Biochemistry and Biophysics* 295 (1992) 297–301.

[36] G.M. Dhavan, The integration host factor-DNA complex: Base-pair opening dynamics measured by NMR and kinetics of complex formation. Ph. D. thesis, Yale University, 1999.

[37] T. Ruusala, D.M. Crothers, Sliding and intermolecular transfer of the lac repressor: Kinetic perturbation of a reaction intermediate by a distant DNA sequence. *Proc. Natl. Acad. Sci. USA* 89 (1992) 4903–4907.

[38] K.M. Parkhurst, R.M. Richards, M. Brenowitz, L.J. Parkhurst, Intermediate species possessing bent DNA are present along the pathway to formation of a final TBP-TATA complex. *J. Mol. Biol.* 289 (1999) 1327–1341.

[39] K.A. Johnson, Rapid quench kinetic analysis of polymerases, adenosinetriphosphatases, and enzyme intermediates. *Meth. Enzymol.* 249 (1995) 38–61.

[40] R.B. Winter, O.G. Berg, P.H. von Hippel, Diffusion-driven mechanisms of protein translocation on nucleic acids. 3. The Escherichia coli lac repressor-operator interaction: Kinetic measurements and conclusions. *Biochemistry* 20 (1981) 6961–6977.

Advances in
DNA Sequence-specific Agents
Series Editor: Graham B. Jones
URL: http://www.elsevier.nl/locate/series/adna

Aims and Scope:

Advances in DNA Sequence-specific Agents (ADNA) is intended to give the reader an up-to-date view of both established and emergent trends, in research involving DNA-interactive agents with an emphasis on sequence specificity. DNA sequence specificity plays a critical role in a number of biological processes, and influences a diverse range of molecular recognition phenomena, including protein-DNA, oligomer-DNA and ligand-DNA interactions. This series encompasses design, synthesis, application, and analytical methods (including clinical and in vitro) for the study of these critical interactions. Recent topics that have been addressed include the mechanisms of selective DNA-topoisomerase poisoning by anti tumor agents, sequence-specific recognition of DNA by groove-binding drugs and drug-conjugates, DNA-cleaving anti tumor chromoproteins, application of sequence-specific anti sense and anti-gene therapy in oncology, and mimetics of the DNA structure. As our understanding of the genome and proteome expands, general developments in the field of DNA sequence-specific interactions are likely to play an increasingly important role. Key developments are likely to involve small molecules with highly specific groove binding, molecules with affinity for unique DNA micro environments, oligonucleotides capable of stabilizing unique DNA architectures, macromolecules which form unique ternary complexes, and molecules which influence the recruitment and assembly of transcription factors. Accordingly, manuscripts have been solicited from experts covering a diverse range of scientific backgrounds, reflecting the cross-disciplinary and dynamic nature of the series.

Books Published:

1. L.H. Hurley, Advances in DNA Sequence-specific Agents, Volume 1, 1992, 1-55938-165-5
2. J.B. Chaires, Advances in DNA Sequence-specific Agents, Volume 2, 1996, 1-55938-166-3
3. M. Palumbo, Advances in DNA Sequence-specific Agents, Volume 3, 1998, 0-7623-0203-8
4. B.J. Chapman, Advances in DNA Sequence-specific Agents, Volume 4, 2002, 0-444-51096-6

Printed and bound by CPI Group (UK) Ltd, Croydon, CR0 4YY

03/10/2024

01040419-0015